AF702204

Springer Texts in Education

Springer Texts in Education delivers high-quality instructional content for graduates and advanced graduates in all areas of Education and Educational Research. The textbook series is comprised of self-contained books with a broad and comprehensive coverage that are suitable for class as well as for individual self-study. All texts are authored by established experts in their fields and offer a solid methodological background, accompanied by pedagogical materials to serve students such as practical examples, exercises, case studies etc. Textbooks published in the Springer Texts in Education series are addressed to graduate and advanced graduate students, but also to researchers as important resources for their education, knowledge and teaching. Please contact Yoka Janssen at Yoka.Janssen@springer.com or your regular editorial contact person for queries or to submit your book proposal.

Festo Kayima

Preparing for Chemistry Teaching

Enhancing Didactic Knowledge & Skills

Springer

Festo Kayima
Department of Teacher Education
Norwegian University of Science
and Technology
Trondheim, Norway

ISSN 2366-7672 ISSN 2366-7680 (electronic)
Springer Texts in Education
ISBN 978-3-031-97121-1 ISBN 978-3-031-97122-8 (eBook)
https://doi.org/10.1007/978-3-031-97122-8

This Springer imprint is published by the registered company Springer Nature Switzerland AG
The registered company address is: Gewerbestrasse 11, 6330 Cham, Switzerland

Contents

List of Figures

List of Tables

1 The Nature of Chemical Knowledge

1.1 Chemistry for Chemistry Teachers

Training to become a chemistry teacher entails acquiring sufficient knowledge of chemistry (chemistry content knowledge) as well as the knowledge, skills and abilities with which to package, repackage, and present chemistry concepts (contents) in a form which is consumable (fathomable and understandable) by target students. This chapter considers the important features of chemistry as a body of knowledge, features which inform and shape how chemistry teachers (including trainees) should think about chemistry as a school subject, plan and implement chemistry teaching. Three questions are considered, (i) what counts as chemical knowledge? (ii) How does this knowledge come into being? and (iii) what is the place and function of chemical knowledge in society? First, let us clarify what we mean in this chapter when we use the term "chemical knowledge" to avoid any confusion that might arise. Some science education scholars (educators) use the term "chemical knowledge" as an umbrella term, and describe at least three kinds of chemical knowledge, namely, (i) the ontological chemical knowledge, (ii) epistemological and (iii) ethical chemical knowledge (see for example; Çetin et al., 2010; Erduran & Kaya, 2019b; Eriksen, 2002; Vesterinen & Aksela, 2009; Vesterinen et al., 2009). Accordingly, the ontological chemical knowledge represents the contents that comprise and distinguish chemistry from the contents of other natural sciences (physics, biology, astronomy, etc.). That is, chemistry—knowledge about the composition and properties of matter, the changes it undergoes and why (Gibney, 2018). The epistemological knowledge of chemistry entails the historical, philosophical and sociological perspectives on chemical practices, while ethical knowledge is concerned with reflections about the role of chemistry and chemistry education in society (Eriksen, 2002). In other words, according to these scholars, chemical knowledge constitutes and accounts for all questions that concerns chemistry, ranging from the contents of chemistry (concepts, laws, theories, principles and models), the ways, practices, processes and the inquiry that result

F. Kayima, *Preparing for Chemistry Teaching*, Springer Texts in Education,
https://doi.org/10.1007/978-3-031-97122-8_1

in chemistry knowledge, including the history of chemistry, philosophical, sociological, and ethical perspectives underlying practices in chemistry and its role in society (e.g.; Erduran & Kaya, 2019b).

There are those scholars on the other hand who use the term "chemical knowledge" to refer only to the ontological part of chemistry, that is, the body of knowledge and understandings developed over time about matter, its composition, structure, properties and the changes it undergoes. These attempt to lay a clear distinction between (i) what counts as chemical knowledge, (ii) the ways, practices, processes and the inquiry that result in this chemical knowledge, and (iii) the function and application of this knowledge in society. This latter and simplified use of the term chemical knowledge allow us to distinguish between the products of the work of chemists and chemistry research endeavors, which indeed are chemistry concepts, laws, theories, models and principles, and the nature of chemistry (NOC) which should describe what chemistry is, how it works (epistemic practices—Erduran and Kaya (2019b)), how chemists operate, and the function and role of chemistry in society. The use of NOC (nature of chemistry) instead of NOS (nature of science) is intended to highlight the domain-specificity of the nature of the knowledge of chemistry which has increasingly become more clearer with continued developments from the philosophy of chemistry research enterprise (Erduran & Kaya, 2019b; Erduran & Scerri, 2002). Nevertheless, chemistry is part of NOS and questions that concern NOS also concern chemistry and vice versa except where and when issues to do with subject domain-specificity are invoked.

Going back to the three questions stated at the beginning of the chapter and considering the distinctions presented above, it is both reasonable and justifiable to conclude as summarized in Fig. 1.1 that the kind of knowledge of chemistry for chemistry teachers comprises three dimensions, (i) the chemical—ontological knowledge, (ii) knowledge of the nature of chemistry, and (iii) knowledge about the interactions between chemistry and society. This classification is easier to conceptualize and resonates especially when the interest is to link the teaching and learning of chemistry with the chemistry curriculum and its development and the role of chemistry in society. Erduran and Kaya (2019b) characterize the different components of a body of the knowledge of chemistry using the term "epistemic core". They describe the epistemic core of chemistry to represent a set of epistemic ideas that underpin how chemistry knowledge production and development occur. Accordingly, the epistemic core of chemistry comprises *(i) aims and values of chemistry*—that is, why and how chemical knowledge is pursued (e.g. goals of chemistry knowledge and chemistry research endeavors and ethical considerations), *(ii) practices in chemistry*—ways in which chemists derive at or construct knowledge, that is, the activities chemists engage in such as experimentation, modelling, etc., (iii) methods in chemistry—approaches chemists use to generate and justify knowledge such as observational and experimental methods, including issues of reliability and credibility of chemical findings, and (iv) knowledge in chemistry—a framework that illuminates that chemistry research or investigation endeavors result in different kinds of knowledge which are structured and interconnected (e.g., chemical concepts theories, laws, and models).

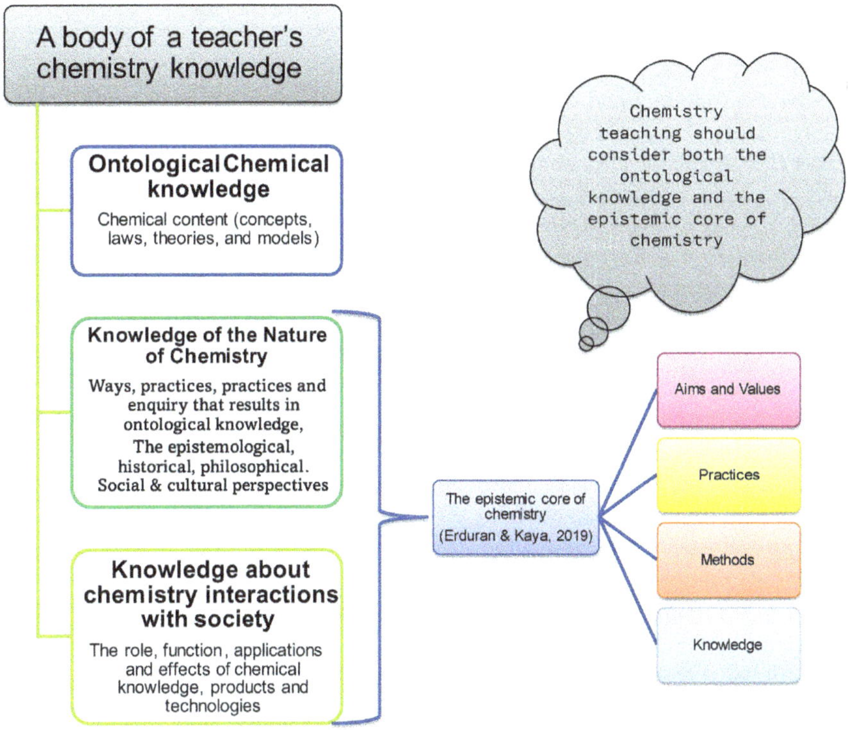

Fig. 1.1 Constituents of a body of knowledge of chemistry for a chemistry teacher

Teachers, educators and students can gain a full understanding of the domain-specificity of chemistry when they understand the different epistemic ideas that underpin how chemistry knowledge production and development occur. Both students and teachers of chemistry need to understand and distinguish what counts as chemical knowledge, the historical, epistemological and social and cultural perspectives underlying the development of this knowledge from the nature of knowledge of other natural sciences. Epistemic foundations of science account for the nature of science in general (Erduran et al., 2014), however the domain-specificity of chemistry is characteristic of chemistry specific epistemic features, that is, the epistemic core of chemistry (Erduran & Kaya, 2019b), which should be explored and understood by both teachers and students of chemistry. Perspectives from philosophy of chemistry provide insights into the nature of chemical knowledge by exploring the concepts, theories, and methods of chemistry, and comparing these with the concepts, theories and methods of other special sciences such as physics and biology (e.g.:, Erduran & Scerri, 2002; Labarca & Lombardi, 2007). Moreover, several benefits to students and teachers are provided in chemical education literature (e.g.; Erduran & Kaya, 2019b; Sjöström, 2013) for why perspectives from philosophy of chemistry should be integrated into chemistry teaching. Examples include both teachers and students gaining a profound comprehension of how chemical knowledge is constructed, validated and applied through exploring epistemic foundations of chemistry as they engage in chemistry learning, as well as promoting and enhancing epistemic practices such as critical thinking, argumentation and engagement in scientific reasoning and evaluation.

As chemical educators continue to advocate for integrating epistemic foundations of chemistry (nature of chemistry—including the historical and epistemological aspects of chemistry) into chemistry teaching, there are notable obstacles to this effort, also well documented in literature, which we must acknowledge and understand their nature to be able to find a way around them. First, integrating epistemic foundations of chemistry into chemistry teaching implies adjusting our teaching priorities from focusing only on factual knowledge and procedural skills and then adopt a teaching that considers epistemic foundations of chemistry as part of what students should learn in and about chemistry. Erduran and Kaya (2019b) provide an example of an acid-base chemistry activity where students explore everyday examples of acids and bases and consider properties distinguishing acids and bases. In addition to this well-known practice of introducing and teaching acids and bases, the authors argue that a teacher should provide for students to explore and understand why acids and bases have such properties. This further opening up- is enriching the lesson with an epistemic purpose according to Erduran and Kaya (2019b). However, it demands adjusting how the lesson is instructionally organized along with re-defining the learning outcomes. Note that it is not only a matter of refocusing our teaching priorities since what is taught in chemistry is governed by what is included in curricula documents. If epistemic themes of chemistry are not considered in the curriculum, then it becomes difficult for teachers to consider for example the nature of chemistry in their teaching plan. The lack of epistemic themes in curricula thus, is a second obstacle that teachers will need to address. When we talk about teachers learning to teach chemistry, one of the expectations is for teachers to understand what is important for students to learn about and in chemistry and how they should learn this. It is thus the responsibility of the teacher to design a teaching that lays the foundation for students to understand chemistry, the nature of this knowledge and its applications. It is therefore possible to work around the obstacle of having epistemic themes lacking in the curriculum by the teachers themselves considering these aspects as a basis for understanding the ontological knowledge dimension of chemistry.

The third obstacle concerns the teacher's competence. Integrating epistemic foundations of chemistry demands that the teacher has the relevant knowledge and skills covering both content and pedagogical needs. One cannot teach or facilitate someone to learn something they themselves do not have a full understanding of. Research indicates that perspectives from philosophy of chemistry—the historical and epistemological perspectives on chemistry, nature and practices in chemistry have been often left out in past in teacher education courses for chemistry teachers. This is also supported by the fact that the work on philosophy of chemistry and empirical research on how perspectives from philosophy of chemistry can be integrated in chemistry teaching has been limited in the past (see a summary by Erduran & Kaya, 2019b). This trend, however, seems to be changing given the increasing number of chemistry educators advocating for inclusion of these aspects in teacher education. Nevertheless, it is most likely that many chemistry teachers do not have a good knowledge of the nature of chemistry, the historical and epistemological perspectives and hence are not well prepared to integrate epistemic perspectives into their teaching.

This introductory chapter on the nature of chemical knowledge only considers a few aspects of the different knowledge dimensions of a body of chemistry knowledge for chemistry teachers. It is meant to give the reader, you, the teacher or educator a starting point for searching to expand your knowledge about the epistemic foundations of chemistry. The question of how to integrate epistemic perspectives becomes easy once you have sufficient content knowledge. Towards the end of the chapter, the question of how to plan and implement chemistry lessons which consider other dimensions of chemical knowledge along with the ontological dimension of chemistry is discussed with illustrative examples about how to achieve this. Moreover, as you will find, several chapters of this book are dedicated to empowering you with the relevant pedagogical competencies that you will need to plan and execute any form of teaching (teaching orientation or philosophy) that you will choose to adopt or embrace.

1.2 The Nature Of Chemical Knowledge

Chemical knowledge represents a body of knowledge and understandings as the resulting product of many years of chemists' engagement in the exploration of the natural world. It comprises the chemical concepts, laws, theories, models and principles which have been developed over time about matter, its composition, structure, properties and behavior. This is what mainly constitutes what chemistry teachers teach in classrooms or what is found in school chemistry curricula. For a chemistry teacher, it is important to understand what nature of knowledge this is, that you take to your students and how this knowledge is derived. There are several features describing the nature of chemical knowledge, including, chemical knowledge being of a nomothetic character, objective, predictive and heavily theoretical (Taber, 2019), chemical knowledge being mainly conceptual and abstract, systematic and structured, dynamic and constantly evolving, and chemical knowledge, while being reliable and durable, is tentative. Moreover, the interdisciplinarity of chemical knowledge, its qualitative and quantitative characters cannot be overlooked, as well as its social and ethical character (see for example; Erduran & Scerri, 2002; Holbrook, 2005; Kampourakis, 2018; Niaz, 2016; Vesterinen & Aksela, 2009). Knowledge about these different features characterizing the nature of chemical knowledge directly shapes a chemistry teacher's thinking regarding how chemistry teaching should be planned and implemented. To allow for a deepened understanding, some of these features will be discussed in the parts that follow.

1.2.1 Objective: Grounded in Empirical Evidence Obtained Through Experiments and Observations; Reliable and Durable, but Tentative and Evolving

Objectivity is a core feature of the nature of science. The underlying gist is that the process that results in science knowledge should be free from any form of influence or bias. Objectivity is a prerequisite condition for any scientific inquiry and for its resulting results to be valued and accepted by the science community. Chemical knowledge is considered objective knowledge. The objectivity lies in the expectation that the developed and accumulated knowledge and understandings that constitute chemical knowledge, have been developed through a process that is independent of a particular observer or investigator, to the extent that such knowledge remains unchanged if a different observer did the job of observing (performed the same experiment) within the same framework conditions (Taber, 2019). Thus, objectivity goes hand in hand with methods and persons involved in the research process or the investigation. In other words, chemical knowledge is grounded in objective evidence obtained through experiments and observations regardless of the person carrying out the experiments or observations. This person, however, as also noted by others e.g., Taber (2019), must have the necessary and relevant knowledge and skills matching the demands of the investigation in question or otherwise similar results or conclusions cannot expected.

Moreover, it is also worth noting that objectivity in science and hence the objectivity of the results of scientific investigations has been at one point in history a consensus question for a community of researchers to answer. There are existing accounts of instances where a community of researchers have interpreted empirical evidence in a particular way, even when there were alternative interpretations that existed. The consequence of this was the strong resistance to taking up and accepting revelations of powerful and novel discoveries which did not seem to appeal to the interpretive framework of the community of researchers (scholars) at the time. As such, there has been a growing agreement over the years and support among scholars and scientists for qualifying objectivity in science on the basis of a social process of competitive cross-validation through peer evaluation (Niaz, 2016, p. 47). This perspective does not in any way compromise the foundational requirements of objectivity in scientific endeavors. It is rather a shift towards collective responsibility and improved accountability.

Chemical knowledge as is the case with other natural sciences is dynamic and evolving. As chemists continue with their exploratory work, new discoveries are made, existing theories are refined, while some others are replaced. The scientific endeavor is about developing knowledge about the natural world. This work is continuous and involves a continued critical appraisal of the existing understandings (already established knowledge) as the basis for building further and coherent understandings. History has shown that this continued critical assessment of scientific endeavors has often resulted in theories, models and explanations with more explanatory power (Kampourakis, 2018; Niaz, 2016). Tentativeness is thus an important feature of scientific knowledge. Scientific knowledge, while reliable and durable, bears with it a certain residual of uncertainty. Moreover, this uncertainty is also a driving force behind improved understandings of the natural world that we have witnessed over several decades of the scientific research business.

Several lived case examples can be considered to highlight the tentative nature of chemical knowledge, its reliability and durability. One such example is the case of understanding the atom—the building block of matter, its structure and composition. Formal recorded accounts about the possible building blocks of matter are traced back to the fifth century BC, with two competing schools of thought, one from Leucippus and his student Democritus who argued that matter comprised small, fine particles that they termed *"atomos"*, and the second from Aristotle and other philosophers who claimed that matter was a combination of four elements, fire, earth, air and water. Historical records indicate that Aristotle's view dominated that of Leucippus and Democritus' for over 2000 years until the early 1800s when John Dalton provided convincing evidence in support of Leucippus and Democritus's conceptions about the composition of matter. Efforts to establish both the existence and composition of atoms as building blocks of matter continued and there were contributions from different committed individuals, who during their time were either exploring the topic directly, or were working on something completely different, but whose discoveries contributed to the work of understanding the structure of the atom. Some of these included, Nicolaus Copernicus (1473–1543), Francis Bacon (1561–1626), Johannes Kepler (1571–1630), Galileo Galilei (1564–1642), Robert Boyle (1627–1691), Isaac Newton (1642–1727), Albert Einstein (1879–1955), Jean Perrin (1870–1942), Antoine Lavoisier (1743–1794), Joseph Proust (1754–1826), J. J. Thomson (1856–1940), Robert Millikan (1868–1953), Henri Becquerel (1852–1908), Marie Curie (1867–1934), James Chadwick (1891–1974),Ernest Rutherford (1871–1937), Niels Bohr (1885–1962), and Erwin Schrödinger (1887–1961), to mention but a few. Historical accounts of the contributions of each one of the individuals above are available and can be easily accessed. However, our interest here is to use this historical account as a manifestation of the tentative nature of chemical knowledge.

Picking on one episode from the historical development of the modern atomic theory, the work leading to the discovery of the atom as the building unit of matter had been a success by the early 1900s. Moreover, the constituents of the atom and their corresponding charges had been established. For example, the charge of the electron had been established and confirmed through Millikan's Oil Drop experiment, and since the atom had been established to be neutral, it was postulated that there was an equivalent number of particles within an atom that counteracted the negative charge to ensure a neutral atom. The question then that needed to be answered at the time was thus about how the positive particles within an atom were arranged. J.J Thomson proposed around 1904 an explanation and a model (the plum-pudding model) of how the electrons and protons where distributed within the atom. According to his model, the positive charge was continuously distributed throughout the atom while the negatively charged electrons were distributed in the atom like raisins inside a cake or like plums in a pudding (Hentschel, 2009). Thomson's conceptions of the structure of the atom were not superseded until 1911 when Rutherford and his colleagues proposed the nuclear model of the atom, following an experiment by Rutherford in 1909 that resulted in results diverting completely from Thomson's conceptions. According to Rutherford's nuclear model, electrons revolved in empty space around a positively charged nucleus and the nucleus contained most of the atom's mass (Podgoršak, 2006). Niels later around 1913, refined Rutherford's conceptions and suggested the Bohr-Rutherford atomic model, providing an explanation of the atomic structure that superseded all ideas and models before. The work of improving our understanding of the atomic structure continued since not all atomic model problems had been solved by Bohrs' model. In 1926, Erwin Schrödinger introduced the quantum model of the atoms through the exploration of the characteristic feature, the wave-particle duality of small particles like electrons. The complete picture of the constituents of the atom was not complete until 1932 when James Chadwick (1891–1974) discovered the existence of neutrons.

The brief historical account above shows how knowledge about the structure of the atom evolved over time and that the tentative character of science knowledge keeps the ball rolling. It portrays indeed that science, and our understanding of the natural world is continuously evolving and improving. Tentativeness of science knowledge is thus a feature that strengthens the reliability of this knowledge. Even when scientific knowledge is confirmed on account of a social process of competitive cross-validation through peer evaluation (Niaz, 2016, p. 47), it is always subject to change pending new findings and or through a reinterpretation of findings (Lederman et al., 2002; Osborne et al., 2003).

1.2.2 Nomothetic, Predictive and Heavily Theoretical

In their scientific endeavors and as they seek to understand the natural world, chemists mainly seek to identify commonalities, patterns, laws and principles that apply across cases rather than single aspects. Chemical knowledge is thus characteristically nomothetic (Taber, 2019). Scholars from philosophy of chemistry describe a large part of chemical knowledge to be concerned with the qualitative aspects of matter and the use of class concepts (e.g., acids, bases, salts, element, etc.) to represent explored phenomena (Erduran & Scerri, 2002). Such concept classifications are enabled due to the nomothetic nature of the work of chemists. Taber (2019) illustrates the nomothetic character of chemical knowledge by giving an illustrative example of a chemist who might study a crystal of copper sulphate with the goal of producing generalizable knowledge, that is, knowledge that also applies to other copper sulphate crystals. There are several other examples.

For example, if you were asked to distinguish between acids and bases, there are already established features that are unique to bases and those that are unique to acids. Moreover, chemists have structured the chemistry of hydrocarbons for example by use of functional groups, where each function group features different properties unique to that group, for example, alkanes, alkenes, alkynes, alcohols, organic acids, aldehydes, ketones, aromatic vs aliphatic compounds, etc.

Nature and organization of chemical knowledge offers predictive power. Chemical knowledge is grounded in and hence informed by well-established laws and theories. For example, considering the historical development of the atomic structure presented in the previous section. The modern atomic theory, the periodic law, theories of bonding and thermodynamics, all these enable prediction by providing chemists with a working framework. Drawing on already established chemical knowledge (principles, laws and theories) enables the prediction of how other substances would behave, their properties and reaction outcomes. Exploring and developing qualitative class concepts is enabled by the chemists' ability to identify essential features that distinguish between phenomena of interest. The terms "natural kinds" and "sameness" were used by Dupré (1981) to describe the existence of objects in classes defined by common properties. Once chemists explore a given phenomenon and manage to generate and confirm the essential properties of the phenomenon, they can rely upon some of these properties to make predictions about other similar phenomena that could be existing and belonging to the same class.

Moreover, chemical knowledge like other science knowledge, while derived through logical argumentation based upon empirical evidence, is largely theoretical (Taber, 2017, 2019). Notice that there is a distinction between scientific knowledge being theory laden and scientific knowledge being largely theoretical. Regarding the former, while we agree from our knowledge of the nature of science that science knowledge and hence chemical knowledge is grounded in objective empirical observations and experiments, such observations and experimental activities are normally influenced by preconceived ideas (Chalmers, 2013). In other words, there are underlying theoretical presuppositions that influence scientists' perceptions, interpretations, or attention, and hence the process and direction considered in producing scientific knowledge (Andersen & Green, 2013; Niaz, 2016). The latter, about chemical knowledge being largely theoretical, concerns the very nature of chemical knowledge. There are several examples we can cite that show that chemical knowledge is indeed largely theoretical. Consider for example concepts such as atoms and molecules. These are very small entities for us to observe with our naked eyes. To understand the behavior, interactions and properties of such very small entities of matter is a work that relies on theoretical models and calculations. Moreover, theoretical frameworks such as quantum mechanics and thermodynamics are central to understanding, predicting and describing complex chemical systems. Chemists and physicists have relied heavily on theoretical mathematical descriptions such as the Schrödinger equation to understand and describe chemical phenomena, for example changes in physical systems over time.

1.2.3 Chemical Knowledge Intersects with Other Natural Sciences and Appeals to Societal Norms and Expectations

The existence and identity of chemistry as an independent scientific discipline has been discussed time and again over several years (Chang, 2015; Hettema, 2012). Indeed, strong claims and arguments have been made in the past against the existence of chemistry as an independent scientific discipline. For instance, Immanuel Kant cited in a book chapter, "Casual laws and the foundations of natural science" by Friedman (1992), expressed that,

> ….. chemistry can be nothing more than a systematic art or experimental doctrine, but never a proper science. For its principles are merely empirical and allow of no presentation a priori in intuition. Consequently, they do not in the least make the principles of chemical appearances conceivable according to their possibility, for they are not susceptible to the application of mathematics. (p. 189)

A novice chemistry teacher or a chemistry teacher not familiar with this ongoing debate, upon reading the above submission, might be left wondering what this is supposed to imply in terms of chemistry being an independent subject. However, such are changing times in the business of training to teach science. It is expected of a teacher of any subject to understand the historical development of a knowledge that he/she teaches or intends to teach to his/her students. In our case of chemistry, this has to do with understanding the nature of chemical knowledge and its place as part of the natural sciences. Kant's submission as quoted above, and which is one of the several such arguments from reductivists' viewpoints, should not be a case for us to worry about as chemistry teachers. The identity of chemistry as an independent discipline had been well established in the period between the eighteenth and nineteenth centuries. Moreover, questioning chemistry's existence as an independent disciplinary resurfaced again following the emergence and success of quantum mechanics when physicists started viewing chemical phenomena as physical phenomena yet to be explored in physics (Labarca, 2011). Consequently, Kant's expression resonates well with the thinking period merging with the success of quantum mechanics. The domain specificity of chemistry cannot be overemphasized. There are several supporting arguments from philosophy of chemistry that chemistry teachers who, when faced with the question of justifying the place of chemistry as an independent subject, could draw on. For example, not all concepts of the chemical world can be reduced to quantum physics and a conceptual reduction is impossible to achieve since concepts such as for bonding, composition and molecular structure can only be formulated on a chemical level (Labarca, 2011; Scerri & McIntyre, 1997; Weininger, 1984).

The reason for addressing aspects that relate to existence of chemistry as an independent science under the discussion about the interdisciplinarity of chemical knowledge, is for the chemistry teacher to understand that a discussion about the interdisciplinary of chemistry exists only when chemistry retains its place as one of the sciences. It is with this understanding that we can map out and understand the contributions of chemical knowledge across other (natural) sciences. Moreover, while discussing how chemistry overlaps and intersects with other sciences—physics, biology, etc., chemistry teachers need and should interest themselves in the ongoing discussion about the domain-specificity of chemical knowledge. It is no surprise thus when many educators are currently recommending and encouraging the teaching and learning of the nature of science for both teachers and students.

Finally, chemical knowledge appeals to societal norms and expectations. The obligation is to society. The essence of exploring to understand the natural world is for us to harmonize our living within the framework that is containable by the natural world itself. We can think about the "natural world" as a world in its natural form (not created by us humans) comprising all living and non-living entities—Flora and fauna, ecosystems, geographical features, natural phenomena, the atmosphere, etc. While chemistry is about exploring the natural world, scientific research endeavors must happen in harmonious interaction with members of the natural world. This is what good science practice is about, that is, responsibility and professionalism (Mehlich et al., 2017). The practices and the inquiry that lead to the development of chemical knowledge demand an understanding of the potential hazards involved and the impact of the products of chemistry research to all members of the natural world. It demands working within an established framework to safeguard the environment and its vulnerable members and to ensure sustainability. Moreover, the research process and the use of the products of chemical research must adhere to established ethical standards.

1.3 Chemistry Interactions with Society and Environment

We now turn our attention to the multifaceted, dynamic and reciprocal interactions between chemistry, society and the environment, and the teachers' role in integrating these interactions as part of classroom instruction. Chemistry (chemical knowledge including chemical technologies) touches and influences every aspect of our existence. The potential benefits and applications of chemical knowledge and chemical technologies are unlimited in the areas of health and medicine, food production (its quality and quantity), energy and water supply, industry, the economy and our everyday life (consumer products). Meanwhile the dangers in the pursuit and application of chemical knowledge are largely evident. The health risks associated with the use of chemical products and in the production process of chemical knowledge and chemical technologies are enormous, e.g., health problems and diseases from exposure to some chemicals, occupational hazards, side effects from drugs, as well as from misuse of chemical drugs (e.g., drug addiction). Moreover, the effect to the environment is undeniably felt and experienced, chemical pollutants from industrial processes, from agricultural runoffs, from chemical waste, the use of persistent organic pollutants, along with greenhouse gases released from chemical processes and chemical products. Indeed, both from what we observe today and what we know from history, the chemical industry, chemical knowledge and chemical technologies have a central role in the development of (conventional, biological and nuclear) weapons that threaten our very existence. The existential threat to all living entities resulting from climate change is increasingly becoming more real, and human activity, including human interaction with chemical knowledge and technologies are at the center of it.

The requirement for all to be chemically literate is no longer a question of debate but rather one of necessity to enable all citizens' participation in shaping and influencing the direction of chemical research and consumption of its products and technologies. Understanding the interactions of chemistry and society is as important for chemists as it is for students, teachers, teacher educators, parents, politicians, journalists, lawyers, businessmen, the entire public, etc. We cannot leave the responsibility of understanding and anticipating the environmental consequences from chemistry research endeavors to chemists alone.

1.3.1 From a Monofaceted to a Multifaceted Chemistry Teaching

There are two important issues for chemistry educators and chemistry teachers to consider, (i) for learners to acquire the relevant chemical knowledge and skills, including knowledge about the nature of this knowledge and how it comes into being (that is, about the nature and methods of chemistry, its history and development) and about the application of chemical knowledge and skills, and (ii) for learners to understand the dangers in the pursuit and application of chemical knowledge, that is, the complex interactions between chemistry and society and the environment. So far, the latter has not received as much attention as the former in terms of characterizing what students should learn about chemistry, or what chemistry teachers should consider as part of their chemistry teaching.

The focus in chemistry classrooms for a long time has been mainly on understanding chemistry concepts, laws, theories, models and principles (disciplinary concepts and ideas), and less on understanding the nature of chemistry, and later alone the societal contexts where chemistry finds its use (Erduran & Kaya, 2019b; Sjöström & Talanquer, 2014). Among the consequences of this is the failure of students to engage in activities and discussions that involve critical reflections about chemistry knowledge, its impact on society and the environment (Driver et al., 1996; Erduran & Kaya, 2019b). Whereas such a failure is often considered at the classroom—learning level, it has a bearing on human practices after school. When chemistry related personal, socio-scientific and ethical questions remain unaddressed in school, society's ability to collectively address the dangers from applying chemical knowledge and technologies is hampered. Moreover, there is an increased likelihood to leave the responsibility of understanding and anticipating the environmental consequences from chemistry research endeavors to chemical scientists and professionals alone. Countering this, requires a meaningful and relevant chemistry teaching, a teaching that connects chemical knowledge, nature of chemical knowledge, its application in society, and the interplay between chemical knowledge, society and environment (Erduran & Kaya, 2019b; Fisher, 2019; Holbrook, 2005; Pedretti & Nazir, 2011; Sjöström & Talanquer, 2014; Sylvia, 2001; Van Heuvelen et al., 2020).

How then should a chemistry teaching that considers interactions of chemistry with society and environment be realized in practice?

To respond to this question, there are different views and teaching perspectives presented by several chemical educators in support of a multifaceted teaching of chemistry. Among these, Sjöström and Talanquer (2014) in their report about humanizing chemistry education, present some conceptualizations and ideas that a chemistry teacher could potentially translate into practice. They draw on three theoretical frameworks suggested by (i) Johnstone (1982, 1993), (ii) Mahaffy (2004), and (iii) Sjöström (2013), to discuss the different aspects of chemistry which teachers could include in the teaching and learning of chemistry in a way that considers the interactions between chemical knowledge, society and the environment. Figure 1.2 shows three advancing conceptualizations of what chemistry content is and should be about as suggested by Johnstone, Mahaffy and Sjöström respectively.

Johnstone's chemistry learning triangle describes the triplet relationship between aspects of chemistry that are observable and can be experienced through our normal senses (*the macro*), and the aspects at molecular level such as the atoms and molecules that we cannot experience with our senses (*the sub-micro*), and the medium with which we communicate and represent the chemical aspects (e.g.; use of symbols and equations)—*the symbolic*. Johnstone's framework thus describes only the disciplinary concepts and ideas of chemistry. Mahaffy (2004) extended Johnstone's planar triangle of chemistry learning levels into a tetrahedron by adding another component termed the *human element*, which according to him represented "the web of human contexts for learning chemistry" (p. 231),—the everyday life and societal aspects of chemistry. Sjöström (2013) developed Mahaffy's (2004) tetrahedron further by dividing the human element into three

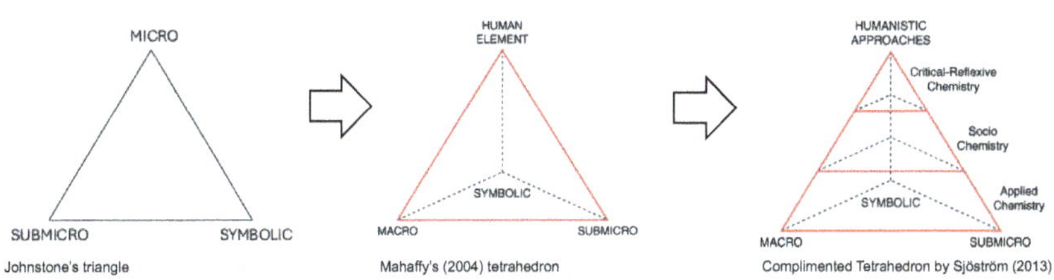

Fig. 1.2 Conceptualizations of chemistry content and the human element of chemistry: Johnstone's triangle reproduced with permission from Johnstone (1993). Copyright (1993), American Chemical Society

levels, (i) applied chemistry, (iii) socio-cultural context, (iii) critical-philosophic approach or critical-reflexive chemistry (Sjöström & Talanquer, 2014).

The "applied chemistry" level entails a focus on the application of chemical knowledge and chemical technologies in our everyday life. Teachers are argued to and are expected to approach chemistry teaching in a way that links the teaching context to corresponding social, ecological and economic implications. A direct link between chemistry taught in classrooms and its applications and implications should be visible for the students through teachers using concrete examples located in our everyday life—in our homes (consumable products), in pharmacies and hospitals, in production and processing industries, etc. The socio-cultural level introduces students to the "development and uses of chemical knowledge, practices and products, as well as understanding the socio-cultural embeddedness of scientific work and ideas"(Sjöström & Talanquer, 2014, p. 1126). In other words, the socio-cultural chemical level addresses the nature of chemistry, the underlying historical, epistemological, sociological perspectives, as well as ethical issues and dilemmas surrounding the pursuit of and application of chemical knowledge and technologies. The third and topmost level of the human element of chemistry concerns the ability to think critically and reflect about chemistry, its nature, process and products and its role in society. It is about being able to and having the basic knowledge about the risks and benefits of chemistry as well as reflecting about the uncertainties in chemistry knowledge generation, the environmental and the human cost involved in the pursuit and application of chemical knowledge and technologies.

Erduran and Kaya (2019a) also provide an empirically grounded example of integrating epistemic aspects of chemistry into chemistry. They describe an intervention developed to "facilitate pre-service teachers' learning to teach nature of chemistry" (p. 85). The intervention comprising 11 3-h sessions was designed to integrate the chemistry's "epistemic core idea", a set of epistemic ideas mentioned earlier, including (i) aims and values of chemistry (ii) chemistry practices, (iii) chemistry methods and (iv) knowledge in chemistry. Various instructional strategies were used or employed in the different sessions including questioning, discussion, presentations, use of visual representations and images, etc. Covered topics in these 11 sessions included introduction to the nature of science, the family resemblance approach (Erduran & Dagher, 2014), aims and values of science, scientific methods, scientific practices, scientific knowledge, and generative images of the epistemic core (pp. 81–102). The instructional organisation of the lessons takes a holistic approach that integrates multiple dimensions to provide a comprehensive learning experience (Borg et al., 2014; Hopkins, 2012; Pauw et al., 2015). A holistic teaching approach is characterized by student-centered learning strategies, visible connection of ideas across science domains (interdisciplinary integration), use of real-world contexts, and situating learning considering the social-emotional and ethical issues. The authors aligned their thinking about how chemistry should be taught and learned with the pedagogical vision of the Next Generation Science Standards (NGSS) that positions three interrelated core science learning pillars, in-depth development of core ideas, focus on science practices, and crosscutting concepts exploring connections between science domains (NGSS-Lead-States, 2013).

Chemistry teachers might perceive a multifaceted chemistry teaching which addresses chemistry disciplinary concepts and ideas, the nature of chemistry and the interactions between chemistry, society and environment, to be too demanding for them. Moreover, formal assessment requirements and curricular limitations could have a substantial influence on what is taught in most chemistry classrooms. For example, the focus on mainly teaching chemistry content (disciplinary concepts and ideas) is often driven by the nature of curriculum expectations and the summative assessment focused on mastery of content knowledge. Under such circumstances, the human element of chemistry is left out completely without any form of accountability. Teachers should understand however that including the human element does not mean planning a separate chemistry teaching. The teaching methodological organization employed and described by Erduran and Kaya (2019a) for example is not different from the way the lesson would be planned when considering the chemistry tetrahedron put forward by Sjöström and Talanquer (2014). Employed strategies such as questioning, discussion, presentations, use of visual representations and images can be considered and included in any other form of instructional lesson organization employing a different didactic framework or model. Consequently, ideas concerning the interactions between chemistry, society and environment can easily be incorporated into the lesson and be addressed alongside chemistry canonical content. Examples of the possible approaches for integrating sustainability issues in chemistry are discussed in Chapter 10. Moreover, the human element provides a strong basis for understanding chemical concepts by linking these to the history behind their development, by linking concepts to the everyday life applications of chemistry, or by reflecting about and discussing the central place of chemistry in shaping society, etc.

1.4 Reflection Questions

1. How would you describe or characterize the kind of chemistry knowledge a chemistry teacher should master?
2. Discuss some of the main features that characterize the nature of chemical knowledge?
3. How does knowledge of the nature of chemical knowledge impact on your thinking about how chemistry should be planned and taught?
4. Describe how important is a teacher's knowledge of the history and philosophy of chemistry regarding planning for chemistry teaching?
5. How would you describe chemistry interactions with society and the environment?
6. Describe your understanding of a multifaceted chemistry teaching using a concrete teaching plan as an example

References

Andersen, H., & Green, S. (2013). Theory-ladenness. In W. Dubitzky, O. Wolkenhauer, K.-H. Cho, & H. Yokota (Eds.), *Encyclopedia of systems biology* (pp. 2165–2167). Springer New York. https://doi.org/10.1007/978-1-4419-9863-7_86

Borg, C., & N., G., O., H. H., & Bergman, E. (2014). Subject- and experience-bound differences in teachers' conceptual understanding of sustainable development. *Environmental Education Research, 20*(4), 526–551. https://doi.org/10.1080/13504622.2013.833584

Çetin, P. S., Erduran, S., & Kaya, E. (2010). Understanding the nature of chemistry and argumentation: The case of pre-service chemistry teachers. *Ahi Evran Üniversitesi Kırşehir Eğitim Fakültesi Dergisi, 11*(4), 41–59.

Chalmers, A. (2013). *What is this thing called science?* (4 ed.). McGraw-Hill Education.

Chang, H. (2015). Reductionism and the relation between chemistry and physics. In T. Arabatzis, J. Renn, & A. Simões (Eds.), *Relocating the history of science: Essays in honor of kostas gavroglu* (pp. 193–209). Springer International Publishing. https://doi.org/10.1007/978-3-319-14553-2_13

Driver, R., Leach, J., Millar, R., & Scott, P. (1996). *Young people's images of science.* McGraw-Hill Education.

Dupré, J. (1981). Natural kinds and biological taxa. *The Philosophical Review, 90*(1), 66–90.

Erduran, S., & Dagher, Z. R. (2014). Family resemblance approach to characterizing science. In S. Erduran & Z. R. Dagher (Eds.), *Reconceptualizing the nature of science for science education: Scientific knowledge, practices and other family categories* (pp. 19–40). Springer Netherlands. https://doi.org/10.1007/978-94-017-9057-4_2

Erduran, S., Dagher, Z. R., Erduran, S., & Dagher, Z. R. (2014). *Reconceptualizing nature of science for science education.* Springer.

Erduran, S., & Kaya, E. (2019a). Incorporating the epistemic core in teacher education practice. In S. Erduran & E. Kaya (Eds.), *Transforming teacher education through the epistemic core of chemistry: empirical evidence and practical strategies* (pp. 81–104). Springer International Publishing. https://doi.org/10.1007/978-3-030-15326-7_4

Erduran, S., & Kaya, E. (2019b). *Transforming teacher education through the epistemic core of chemistry.* Springer.

Erduran, S., & Scerri, E. (2002). The nature of chemical knowledge and chemical education. In *Chemical education: Towards research-based practice* (pp. 7–27). Springer.

Eriksen, K. K. (2002). The Future of tertiary chemical education–A Bildung focus. *HYLE-International Journal for Philosophy of Chemistry, 8*(1), 35–48.

Fisher, M. A. (2019). Systems thinking and educating the heads, hands, and hearts of chemistry majors. *Journal of Chemical Education, 96*(12), 2715–2719. https://doi.org/10.1021/acs.jchemed.9b00346

Friedman, M. (1992). Causal laws and the foundations of natural science. In P. Guyer (Ed.), *The Cambridge companion to Kant* (pp. 3–161). Cambridge University Press.

Gibney, D. (2018). Towards an ideal chemistry curriculum. *School Science Review, 100*(370), 30–35.

Hentschel, K. (2009). Atomic models, J.J. Thomson's "Plum Pudding" model. In D. Greenberger, K. Hentschel, & F. Weinert (Eds.), *Compendium of quantum physics* (pp. 18–21). Springer Berlin Heidelberg. https://doi.org/10.1007/978-3-540-70626-7_9

Hettema, H. (2012). *Reducing chemistry to physics: Limits, models, consequences.*

Holbrook, J. (2005). Making chemistry teaching relevant. *Chemical Education International, 6*(1), 1–12.

Hopkins, C. (2012). Reflections on 20+ Years of ESD. *Journal of Education for Sustainable Development, 6*(1), 21–35. https://doi.org/10.1177/097340821100600108

Johnstone, A. H. (1982). Macro and microchemistry. *School Science Review, 64*(227), 377–379.

Johnstone, A. H. (1993). The development of chemistry teaching: A changing response to changing demand. *Journal of Chemical Education, 70*(9), 701. https://doi.org/10.1021/ed070p701

Kampourakis, K. (2018). Science and Uncertainty. *Science & Education, 27*(9), 829–830. https://doi.org/10.1007/s11191-018-0019-3

Labarca, M. (2011). The relation between chemistry and physics. origin and present relevance of a controversy space. In O. Nudler (Ed.), *Controversy spaces: A model of scientific and philosophical change* (Vol. 10). John Benjamins Publishing.

Labarca, M., & Lombardi, O. (2007). The philosophy of chemistry as a new resource for chemistry education. *Journal of Chemical Education, 84*(1), 187. https://doi.org/10.1021/ed084p187

Lederman, N. G., Abd-El-Khalick, F., Bell, R. L., & Schwartz, R. S. (2002). Views of nature of science questionnaire: Toward valid and meaningful assessment of learners' conceptions of nature of science. *Journal of Research in Science Teaching, 39*(6), 497–521.

Mahaffy, P. (2004). The future shape of chemistry education [10.1039/B4RP90026J]. *Chemistry Education Research and Practice, 5*(3), 229–245. https://doi.org/10.1039/B4RP90026J

Mehlich, J., Moser, F., Van Tiggelen, B., Campanella, L., & Hopf, H. (2017). The ethical and social dimensions of chemistry: Reflections, considerations, and clarifications. *Chemistry—A European Journal, 23*(6), 1210–1218. https://doi.org/10.1002/chem.201605259

NGSS-Lead-States. (2013). *Next generation science standards: For states, by states.* National Academies Press.

Niaz, M. (2016). *Chemistry education and contributions from history and philosophy of science.* Springer.

Osborne, J., Collins, S., Ratcliffe, M., Millar, R., & Duschl, R. (2003). What "ideas-about-science" should be taught in school science? A Delphi study of the expert community. *Journal of Research in Science Teaching, 40*(7), 692–720.

Pauw, J. B.-d., Gericke, N., Olsson, D., & Berglund, T. (2015). The effectiveness of education for sustainable development. *Sustainability, 7*(11), 15693–15717. https://www.mdpi.com/2071-1050/7/11/15693

Pedretti, E., & Nazir, J. (2011). Currents in STSE education: Mapping a complex field, 40 years on. *Science Education, 95*(4), 601–626. https://doi.org/10.1002/sce.20435

Podgoršak, E. B. (2006). Rutherford—Bohr atomic model. In E. B. Podgoršak (Ed.), *Radiation physics for medical physicists* (pp. 42–85). Springer Berlin Heidelberg. https://doi.org/10.1007/3-540-29471-6_2

Scerri, E. R., & McIntyre, L. (1997). The case for the philosophy of chemistry. *Synthese, 111*, 213–232.

Sjöström, J. (2013). Towards Bildung-Oriented chemistry education. *Science & Education, 22*(7), 1873–1890. https://doi.org/10.1007/s11191-011-9401-0

Sjöström, J., & Talanquer, V. (2014). Humanizing chemistry education: From simple contextualization to multifaceted problematization. *Journal of Chemical Education, 91*(8), 1125–1131. https://doi.org/10.1021/ed5000718

Sylvia, W. A. (2001). Teaching chemistry from a societal perspective. *Pure and Applied Chemistry, 73*(7), 1209–1214. https://doi.org/10.1351/pac200173071209

Taber, K. S. (2017). Reflecting the nature of science in science education. In K. S. Taber & B. Akpan (Eds.), *Science education: An international course companion* (pp. 23–37). SensePublishers. https://doi.org/10.1007/978-94-6300-749-8_2

Taber, K. S. (2019). *Foundations for teaching chemistry: Chemical knowledge for teaching.* Routledge.

Van Heuvelen, K. M., Daub, G. W., Hawkins, L. N., Johnson, A. R., Van Ryswyk, H., & Vosburg, D. A. (2020). How do i design a chemical reaction to do useful work? Reinvigorating general chemistry by connecting chemistry and society. *Journal of Chemical Education, 97*(4), 925–933. https://doi.org/10.1021/acs.jchemed.9b00281

Vesterinen, V.-M., & Aksela, M. (2009). A novel course of chemistry as a scientific discipline: How do prospective teachers perceive nature of chemistry through visits to research groups? [10.1039/B908250F]. *Chemistry Education Research and Practice, 10*(2), 132–141. https://doi.org/10.1039/B908250F

Vesterinen, V.-M., Aksela, M., & Sundberg, M. R. (2009). Nature of chemistry in the national frame curricula for upper secondary education in Finland, Norway and Sweden. *NorDina: Nordisk tidsskrift i naturfagdidaktikk, 5*(2), 200–212.

Weininger, S. J. (1984). The molecular structure conundrum: Can classical chemistry be reduced to quantum chemistry? *Journal of Chemical Education, 61*(11), 939.

2 Becoming a Chemistry Teacher

Abstract

This chapter introduces what constitutes the chemistry didactics knowledge a chemistry teacher needs to effectively teach chemistry. It takes up general pedagogical learning theoretical perspectives and draws out connections between these and chemistry teaching and learning. Emphasis is put on existing teacher knowledge characterizations (teacher knowledge elements), their nature, attainment (development) and the envisioned application of these knowledge elements in chemistry classrooms. By describing the various components of a teacher's knowledge for teaching, the chapter aims to enable both teachers and teacher trainees to visualize and appreciate the need for and relevance of chemistry didactics training for chemistry teachers.

2.1 A Chemistry Teacher's Knowledge for Teaching

2.1.1 Didactics or Pedagogics?

The terms "didactics" and "pedagogy" are among the common and frequently used concepts in the professional world of teaching. These terms (or concepts) appear not only to serve as anchoring points upon which the professional language used in educational contexts rests but are also encompassing terms for what constitutes the knowledge, the experiences, as well as all research-based perspectives that inform teaching and learning at all levels. Indeed, as you might have realized already, and depending on which part of the world you are situated in (more particular for those new in this field of education), at least one of these two terms appears in almost all texts that you have encountered so far within the field of teacher education. As such, it is imperative that this chapter, "becoming a chemistry teacher", starts by briefly considering how the education community has conceptualized and operationalized these two terms in varied educational contexts.

F. Kayima, *Preparing for Chemistry Teaching*, Springer Texts in Education,
https://doi.org/10.1007/978-3-031-97122-8_2

The use of the term "didactics" (*Didaktik—from German speaking countries or Didaktikk—for Nordic countries*) in educational contexts is more pronounced in Europe, i.e., in Germany, France and the Nordic countries (Norway, Finland, Sweden, and Denmark). However, in the Anglo-Saxon educational contexts, the term "didactics" did historically bear a negative connotation (Gundem, 2004; Hamilton, 1999) and hence its use in these contexts did not imply the same meaning that was implied when used in Europe. Instead, different descriptions of similar activities that relate to the teacher's knowledge for teaching and research about teaching were being used over the years without explicitly having a well-coined term (Kansanen, 2009). Moreover, the term pedagogy became more acceptable to the Anglo-Saxon educationalists, and its use in educational contexts has been increasing since the early 1970s (Hamilton, 1999). Indeed, the term "pedagogy" is more widely recognized and used compared to the term "didactics" in terms of characterizing ways of teaching, its methods, theory and practices.

Whereas these two terms "didactics" and "pedagogy", each attracts a variety of meanings based on the historical origins of each term, they largely bear similar or close meanings in educational contexts. Some of the more pronounced descriptions provided in educational literature include defining pedagogy as a "science of teaching embodying both curriculum and methodology" (Kansanen, 2009; Simon, 2013), and didactics as "a science and theory about teaching and learning in all circumstances and in all forms" (Gundem, 2004). From these two definitions and also drawing from Gundem's (2004) review, a simple description representing both terms would be to consider pedagogy and didactics as encompassing terms for that which can be taught and learnt, how it could be taught and learnt, and for what purpose or intention. Note that a research component is also reflected (included) in this simplistic definition, given that it is the educational research and the continuity of educational research efforts that feeds into and sustains a teacher's knowledge on and about teaching and learning.

2.1.2 From General Didactics (Pedagogy) to Subject-Specific Didactics

The work of developing and qualifying what counts as a teacher's knowledge for teaching has been ongoing over several decades. Among the widely recognized efforts include the conceptual developments of Lee Shulman around the mid-1970s which saw the classification of teacher knowledge into content knowledge for teachers, subject content knowledge and subject-matter pedagogical knowledge. Shulman later expanded these and came up with seven categories of teacher knowledge (Shulman, 1986, 1987). Following Shulman's work, we now find several classification categories of the teacher's knowledge in educational literature (a summary of these is given by Kayima, 2022). A teacher's knowledge for teaching generally comprises three major knowledge categories, (i) the general pedagogy (didactics), (ii) subject content knowledge, and (iii) the subject pedagogical knowledge (subject-specific didactics). Kansanen (2009), used a Venn diagram (see

Fig. 2.1) to illustrate a logical relationship between these three teacher knowledge elements. General pedagogy (Fig. 2.1) represents a body of "knowledge of principles and strategies of classroom management and organisation that transcend subject matter" (Shulman, 1987, p. 8), or "a specialized knowledge of teachers for creating effective teaching and learning environments for all students independent of subject matter" (OECD, 2017, p. 80). Pedagogical content knowledge, also referred to as "subject-specific didactics" in our case, constitutes the theory, methods, and practice of teaching and learning that specific subject. Shulman described this knowledge as "the special amalgam of content and pedagogy" that is needed for teaching a subject.

Advances in technology have had a direct impact on the educational landscape over the years and thereby shaping the way of conducting educational business, including educational research and the teaching and learning process. A myriad of ever-emerging technological tools and resources with direct or indirect application in education, has meant a shift in our teaching and learning philosophies, unto the need for developing new teaching resources to take advantage of these technologies. Moreover, it also meant rethinking, redefining and expanding a teacher's teaching competence profile described in the above paragraph, to accommodate the use of technology in teaching. In response to this inevitable need, Mishra and Koehler (2006) proposed another category of a teacher's knowledge for teaching, the teacher's Technological pedagogical content knowledge (TPCK), characterizing it as a knowledge that enables a teacher to effectively apply or use technologies in teaching. More details about this kind of knowledge are discussed in Chapter 8, about teaching with technology.

It is reasonable to assume that teacher educators and chemistry teachers already in the field have already experienced the value and or the applicability of both general pedagogy and subject-specific didactics. However, for the majority teacher trainees (student teachers) taking a course in pedagogy or subject didactics, a question often arises regarding the relevance, the need for, and altogether the applicability of what is being studied in these pedagogical or didactic courses. Such questions as, "Do we really need this chemistry didactic course, given that we already have a good mastery of chemistry content?", have always been on the hearts of many student teachers (chemistry teacher trainees) with whom I have interacted over several years of my career in teacher education. Indeed, in some faculties, attendance of such courses has been made compulsory for students, having recognized that their students are not naturally inclined to attend lectures in these courses. You (the teacher trainee) may also have similar or related questions. Well, the view that anyone can teach or implement teaching in a given subject as long as they exhibit sufficient content knowledge is not new. It can be argued in fact that it is this very perceptual understanding that resulted in a strong research movement around the mid-1970s, that was aimed at characterizing a different kind of teacher knowledge that would help teachers achieve classroom learning goals (Lederman & Lederman, 2015; Shulman, 1986). Those involved in teaching and research on teacher effectiveness in the mid-1970s, had realized that being an expert in the subject content did not necessarily translate into being able

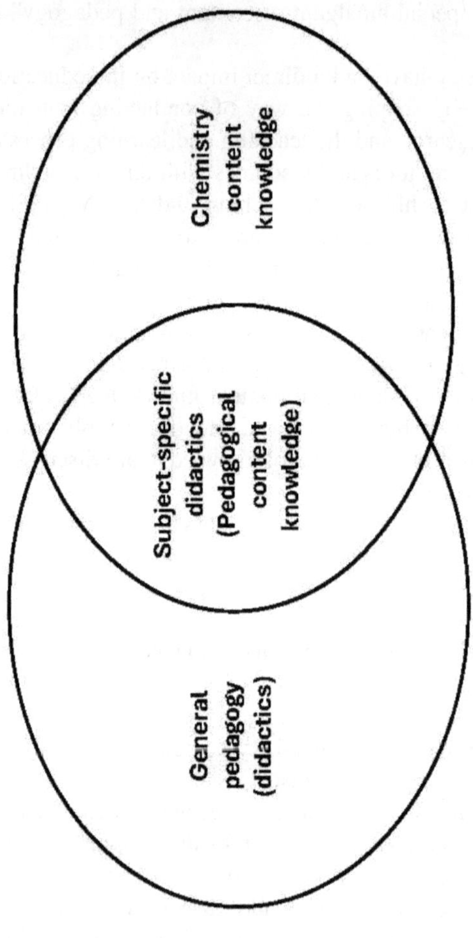

Fig. 2.1 A logical relationship between the three teacher knowledge elements (idea adopted from Kansanen's (2009) study)

to teach effectively. On the contrary, there was needed a different form of knowledge (skills and competences) that informs the teaching, and that which mediates a transformation of subject matter into instruction (Bucat, 2005; Kayima, 2022). In the same way, you will also need a different kind of knowledge in addition to your chemistry content knowledge to be able to plan, implement teaching, as well as monitor and facilitate the attainment of intended learning aims. This is how and why we get to talk about both general pedagogy and pedagogical content knowledge (subject-specific didactics).

Equally important is the recognition that as newcomers (beginners) in the field of teaching, it might be difficult in your current situation to visualize or realize the value of subject pedagogic or didactic courses until you get the opportunity to experience the use of this knowledge in real practice. In fact, for trainees, you will come to realize during your first school practice (first teaching experience) that you indeed need more knowledge categories than the summary showed in Fig. 2.1 to achieve the kind of learning you would desire for your students. Some educators have argued that effective teaching "evolves from a dynamic interplay of generic beliefs, skills and knowledge and subject-specific competences (for details see Ulferts, 2019, p. 8). This conclusion is well supported by the experiences I have had with my students when they come back from their school practice period. Quite often, they raise questions about the need for a different kind of knowledge or competence, with which they can confront and deal with daily classroom challenges that they (have) faced during school practice. This kind of knowledge that you (the teacher-trainee) will also need is a constituent part of the teacher's knowledge for teaching. Moreover, the educational research community has also gone ahead to identify and characterize constituents of this knowledge. For instance, Barnett and Hodson (2001) distinguish this knowledge into the teacher's professional and classroom knowledge, others have used terms such as "teacher's professional craft knowledge" (Brown & McIntyre, 1993), "teacher lore" (Schubert & Ayers, 1992), "action-oriented knowledge" (Carter, 1990), "teacher's personal practical knowledge" (Connelly & Clandinin, 1985), teacher's context-related knowledge (van Driel et al., 1998), etc. A more overarching description for this kind of knowledge is given by Barnett and Hodson (2001), who describe a teacher's professional knowledge as "the knowing of teaching by unconsciously reflected experience" (p. 438), and a teacher's classroom knowledge as a knowledge that enables the teacher to "think on their feet at both a micro and macro level, observe their students and constantly adjust their tone, delivery, activities, verbal interactions, and so on, to ensure that the lesson proceeds as intended." (pp. 438–439).

Note that this other category of knowledge *(teacher's professional and classroom knowledge)*, though greatly influenced by the underlying knowledge of general pedagogy, specific-subject didactics and content knowledge, it is a knowledge that develops in and with practice, and therefore grounded in other societal, cultural and personal experiences, attitudes and beliefs about learning processes. Nevertheless, we can view or consider the teacher's professional and classroom knowledge to be intersecting both the teacher's subject-didactics and general pedagogical knowledge as the illustration shows (Fig. 2.2). As already mentioned, unlike the general pedagogics, the content knowledge, and the main elements of the subject-specific didactics, the teacher's professional and classroom knowledge are knowledge elements largely acquired in and with practice (Barnett & Hodson, 2001; Kayima, 2022). This implies a limitation to how much teacher educators can do to support teacher trainees in acquiring and developing this form of knowledge. Some teacher educators and scholars in recent years have endeavored to find ways to support teachers and teacher-trainees in characterizing and developing this knowledge. Among a few, are Kayima (2022) and Loughran (2002) who emphasize the need to provide teacher-trainees with opportunities to develop their own assertions or personal theories about teaching, as a way that provides for transitions in thinking about the practice of teaching. We will get back to this in Chapter 11, about the role of school practice. What is important for you at this point is to have a good understanding of and an appreciation of the various elements which constitute what is referred to as teacher knowledge, and an overview about how the different elements of teacher knowledge could be developed or acquired.

2.1.3 Chemistry Didactics (Chemistry Pedagogical Knowledge)

In the previous sub-sections of this chapter, we have considered a general overview of the teacher knowledge elements normally understood and presented in the form of theoretical knowledge about teaching. With this overview, let us now turn our attention towards chemistry-subject didactics. To begin with, let us consider and keep in mind the following questions.

1. What exactly is the concern of a chemistry didactics course?
2. What do you expect to obtain in terms of knowledge and skills when you attend a chemistry didactics course?
3. How different is this knowledge from the general pedagogical knowledge or other-subject-specific didactics?

Chemistry education, a disciplinary field concerned with the teaching and learning of chemistry has over the years been concerned with addressing mainly five questions *(i) what is it to be taught (in terms of content)? (ii) How can or will it be taught? (iii) How do students learn (iv) what has been learned or what is learned (by students)? and (v) what new research developments can be adopted to affect what is learned or the process of learning positively?* (for a historical development on this see Bucat, 2004). Emphasis in chemistry didactics courses is majorly put on the first four questions. The "what to teach" question, concerns the chemistry content that is considered relevant, that which is included in the curriculum. Different countries have defined their individual sets of education goals, and it is these goals that define what kind of subject contents or topics are relevant at various levels

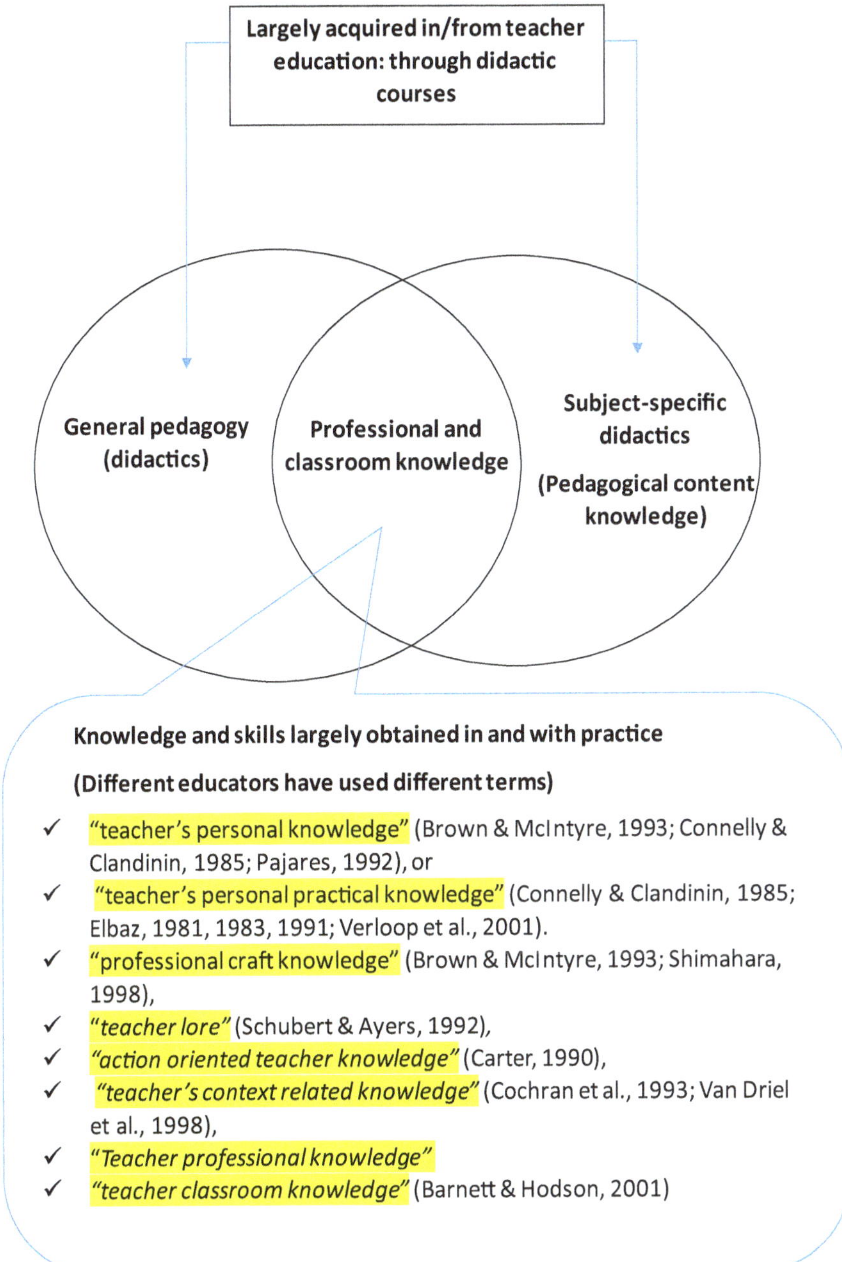

Fig. 2.2 An overview of the teacher's knowledge for teaching acquired with and in practice

of education. The curriculum document is the governing document for you as a teacher, and it defines the working framework for your teaching. Some curricula documents are so strict and control the nature, complexity and scope of content to be taught, while other countries choose to adopt a more flexible approach. We will come back to this aspect later in Chapter 3. The "how it is taught", "how students learn" and "what is learned" questions are a central feature of the subject-didactics knowledge, and by considering these questions further, we will derive at the answers to the three questions that I posed at beginning of this sub-section. To put it in a simpler way, let us consider the chemistry didactics course to comprise four teacher knowledge components, that is, knowledge about "*what to teach,* "*how to teach*", "*how students learn*", *and* the knowledge concerning "*what is learned (by students)*".

2.1.3.1 The "What to Teach", "How to Teach It" and "What is Learned" Teacher Knowledge Components

As expressed earlier, the "what to teach" concerns what chemistry topics are considered relevant to teach at a given classroom level, and curricula documents for different educational contexts often dictate in this matter. Our responsibility as teachers, teacher educators, and or teacher-trainees, is to ensure that we have an excellent mastery of the chemistry content. When a working definition for what counts as the teacher's pedagogical content knowledge was conceived initially (Shulman, 1986, 1987), what mainly distinguished general pedagogy from pedagogical content knowledge was that the latter was conceived as "a knowledge about teaching and learning of particular subject matter that takes into account the learning demands inherent in the subject matter." From this description, we can draw two conclusions: first, there are differences between different subjects when it comes to teaching or studying them. That is, the preparations made, the thinking required, the knowledge and skills needed for teaching chemistry will be different for biology, physics and so on. Second, each topic within chemistry will also vary in terms of what kind of knowledge, skills and preparations a teacher needs and will make, to be able to teach that topic. This characteristic distinction of the inherent differences between subject-specific didactics from general didactics is well recognized. Geddis (1993) noted for example that "beginning teachers need to learn not just "how to teach" but rather "how to teach electricity", "how to teach world history", or "how to teach fractions" (p. 675). Bucat (2004) adds to Geddis' submission in his attempt to characterize chemistry didactics knowledge for teachers. That is, teachers need to learn "how to teach stoichiometry" or how to teach chemical equilibrium, and so on .... (p. 217). The central idea in these two expressions is about the teacher's ability to re-package chemistry content knowledge and present it in a way that is commensurate with the level of the learners in question. This demands that a teacher not only has sufficient knowledge about a particular chemistry topic but must also have a knowledge that includes multiple aspects about the topic which is helpful when it comes to planning and teaching the topic. That is to say, the "knowledge of subject matter, its teachability and learnability" (Bucat, 2004; Geddis, 1993).

From this background, you realize that the *"how to teach" component* is very much closely related to the *"what to teach" component*. As chemistry teachers, we can only start to develop sound strategies to teach a certain topic only if we have a deep understanding of that topic. On the contrary, it will be very easy for us to simply give one-way communication lectures if we are not confident about our subject matter knowledge (what to teach). Moreover, even if we have acquired sufficient general pedagogical knowledge and skills, our creativity in classrooms will be hampered if we do not command a good understanding of the topic (concept) we are handling at any given moment in our teaching. Consequently, chemistry didactics as a knowledge of and about teaching comprises knowledge elements that feed into one another (Fig. 2.3).

So far, when considering *the how to teach component* of the subject didactics, we have only looked at the difference between having knowledge about the topic and having knowledge about the particular teaching and learning demands of that particular topic. However, this alone is not sufficient for the teacher to be able to plan and conduct teaching of a specific chemistry topic. The *how to teach component* thus extends further to include other knowledge elements, which altogether make it possible for a teacher to plan and execute teaching in varied ways, based on the nature of the content as well as the contextual factors of the teaching–learning environment. Moreover, at the beginning of this sub-section, I mentioned that chemistry education was also concerned with a fourth question that deals with finding better ways of teaching and learning through research and other educational initiatives. Research efforts resulting from this have greatly impacted on what teacher knowledge elements are considered relevant for us as chemistry teachers. The *how to teach component* thus comprises several other didactical topics considered a part of the knowledge package that enables classroom teachers to plan and execute lessons using a variety of techniques and approaches among other things. Indeed, it is these important topics that together form a compendium of the chemistry didactic knowledge largely covered in a chemistry didactics course. This entire book addresses most of these topics in the respective chapters as you will find.

The focus of the *"what is learned (by students)"* component is on the intended students learning outcomes and how we get to establish the achievement of these. Underlying this dimension is the concept of assessment that will be considered in Chapter 6. Having defined what to teach (content and intended learning outcomes) and how to teach it (with what methods, tools/aids and teaching techniques), there is the need to establish the attainment of whatever was set out to be taught and questions about assessment and how to effectively use various assessment strategies to affect chemistry learning are addressed in Chapter 6.

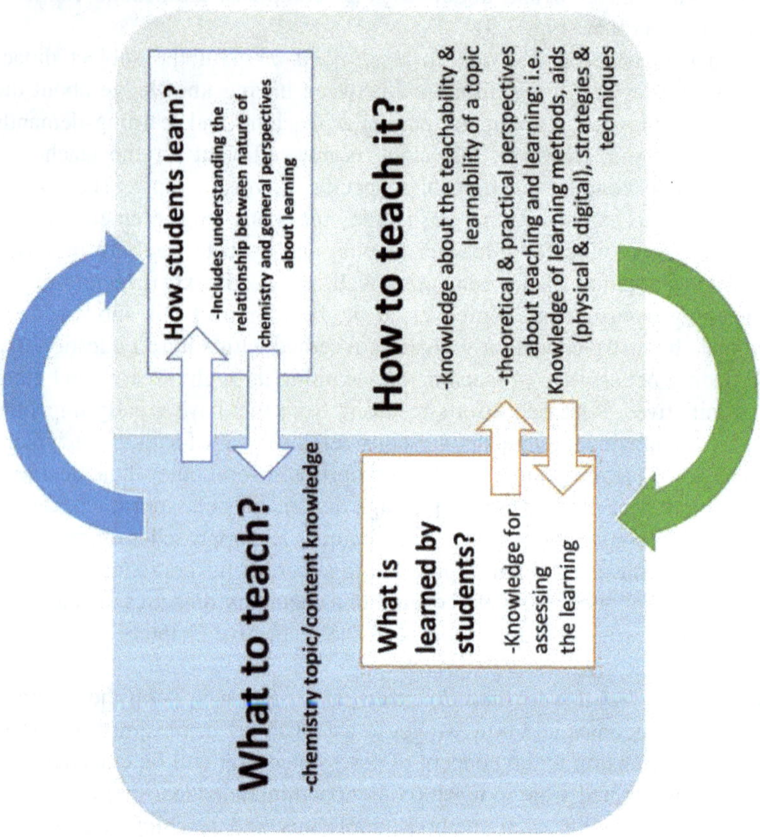

Fig. 2.3 The four-knowledge dimension that constitute subject-specific didactics

2.2 Theoretical Perspectives that Underpin Chemistry Teaching and Learning

So far, the breakdown of a chemistry teacher's knowledge for teaching in the above subsections has included only a teacher's knowledge about *what to teach, how to teach it*, and *what is learned* by students. Another important sub-component of the teacher's knowledge base that feeds into the teacher's *how to teach knowledge dimension*, concerns a teacher's knowledge about *how students learn* (Fig. 2.3). A range of different educational (learning) theories have been suggested throughout the twentieth century to account for students' learning. Three well-known learning theories, that you most probably will encounter in general pedagogy courses include behaviorism, cognitivists and participationist theories—e.g.; sociocultural theories, situated cognition, cultural-historical activity theory, and social constructivism (Chinn & Iordanou, 2023; Danish & Gresalfi, 2018). Different assumptions about learning are advanced in each theory, and teachers can draw on one or a combination of these ideas to make decisions about the most appropriate strategies, tools and methods for specific topics in different learning situations or contexts. While there is currently more pronounced support for constructivists' (mostly social constructivists') learning views in science education (Agarkar & Brock, 2017; Cooper & Stowe, 2018), other theories also provide useful insights that you will find applicable to teaching situations. For instance, you will soon discover (if not already discovered) that behaviorists' ideas of conditioning, reinforcement, and repetition, as well as views about packaging content into small manageable units to enable learners to receive, store and process information (cognitivists' views), directly or indirectly support the operationalization of constructivist approaches to teaching.

We will not go into the details of these respective theories about learning, however, teacher educators using this book will be expected to elaborate more on these theories or even take up more perspectives about how students learn depending on varied needs of their trainees (or the guiding curriculum). For those classroom teachers or teacher trainees who desire a more detailed insight into the various general theories about learning, brief summaries of these and their historical developments are provided by among others, Agarkar and Brock (2017); Chinn and Iordanou (2023); Ertmer and Newby (2013); Leach and Scott (2003) and Cooper and Stowe (2018).

That said, let us turn attention to the question of *what it means to "know how students learn" in the context of teaching chemistry*. The uniqueness of chemistry, the perceived nature and complexity of its contents, require a closer consideration of how our students (learners) will construct an understanding of chemistry principles, or what could be the possible hindrances to achieving this construction. If you (the chemistry teacher, educator or trainee) are to consider and draw on conceptions advanced in different learning theories, to plan and teach chemistry, you will need to also understand the particular challenges inherent in making students understand a subject, whose nature comprises both intuitively accessible and less accessible phenomena. In other words, effective teaching of chemistry relies heavily on understanding the "nature of chemistry" itself. Understanding the "nature of chemistry" should be distinguished from having a good knowledge of chemistry content or specific topics of chemistry. Rather, it concerns understanding *what characterizes chemical knowledge? How this knowledge comes into form (history and philosophy of its development)? And what the overall goal of chemistry is.* A brief introduction about the nature of chemistry has been provided in Chapter 1.

As part of the "nature of science", chemistry aims to increase our understanding of the structure and behavior of the matter of our environment. That is, "chemistry is—the science that seeks to understand the behavior of matter by studying the behavior of atoms and molecules." (Tro, 2017, p. 40). As a subject, chemistry does constitute "two distinct levels of formal concepts that need to be related to each other and to observed phenomena" (Taber, 2013, p. 165). We can think about the process of teaching and learning of chemistry as one of navigating two worlds, the molecular *(-atomic)* world—the less intuitively accessible world, and the macro world—the physically experienced world (matter, *including air the less obvious component of matter*). Moreover as Cooper and Stowe (2018); Johnstone (1991); Taber (2013), among other educators express that, a specialized language comprising not only technical vocabulary, but also symbolic forms of representation, is used when communicating various understandings at the two worlds/levels.

Teaching challenges that relate directly to the nature of chemistry itself have been highlighted in chemistry education research, including a synthesis of ideas about how students' learning of chemistry can be realized. First, chemistry experts describe chemistry at "various levels of thought" (Cooper & Stowe, 2018; Taber, 2013). One conception that has greatly influenced how chemistry educationalists approach the question of *how students should learn* chemistry, is Johnstone's three conceptual levels of chemistry (Fig. 2.4), the *macro, Sub-micro* and *the symbolic (representation)* levels of chemistry (Johnstone, 1982, 1991, 1993).

There has been several reviews and discussions about the objectivity and hence acceptability of Johnstone's three conceptual levels of chemistry in chemistry education research (see for example; Cooper & Stowe, 2018; Taber, 2013). Taber (2013), for instance, looking at the relationship between the three levels, *the macro, sub-micro* and *the symbolic*, concludes that *the symbolic level* "cannot be readily separated" (p. 165) from the other two levels. Rather, the *symbolic level* concerns representing and communicating the concepts, the observations, the understandings, or the models developed at the two levels. Thus, it is *the language dimension level* of chemistry, comprising the languages tools for use in understanding whatever interactions have happened at the two levels. This synthesis by Tabar (2013) (and others who share in his understanding) does not imply a misalignment with Johnstone's ideas, but a step in the forward direction regarding our conceptualization of the "nature of chemistry".

Moreover, an important submission for us teachers from Johnstone's accounts is the argument that a chemistry teaching that demands students to work at or

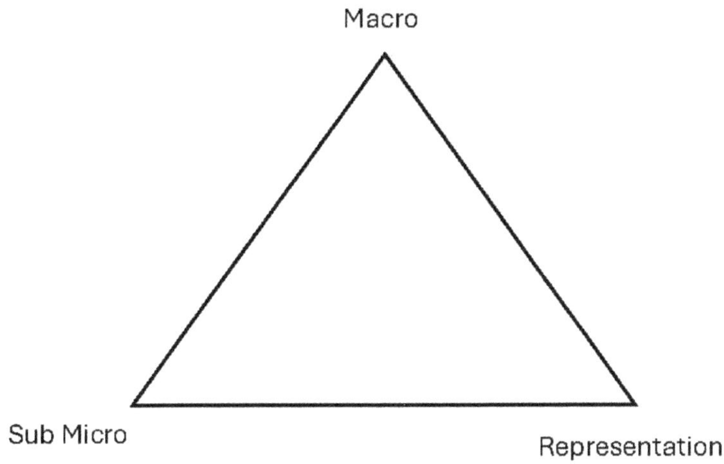

Fig. 2.4 The three conceptual levels of chemistry, reproduced with permission from Johnstone (1993). Copyright (1993), American Chemical Society

produce an understanding of chemical concepts of interest at all three levels simultaneously, presents learning difficulties to learners. Johnstone and others after him, had suggested that teachers should (based on the nature of students at hand) concentrate at one corner rather than teaching at the middle of the triangle. However, this teaching perspective appears (from my point of view) to position the teacher as guiltless (that is, the teacher knows what to do in terms of teaching), and that it is only the students who are implicated in terms of failure to grasp chemical concepts. Undoubtedly, the reasons underlying the difficulty students have when it comes to learning chemistry are (more or less) equally shared between the teacher on one hand and students on the other. That is, teachers themselves need to understand the inherent nature of chemistry and combine this knowledge with conceptions about how individuals learn. As we saw in Chapter 1, chemical knowledge comprises both the ontological chemical knowledge and the epistemic aspects of chemistry (Erduran & Kaya, 2019). Understanding the inherent nature of chemistry is thus not limited to only understanding the three chemistry dimensions representing the ontological chemical knowledge but also involves considering and getting to understand chemistry epistemic aspects, accounting for the historical, philosophical, epistemological, sociological and ethical perspectives underlying practices in chemistry. This teacher's knowledge piece is vital in the planning and packaging of information regardless of whether one chooses to concentrate on one side of Johnstone's triangle or in the middle of it.

Today (July 2025) we find a huge amount of teaching and learning perspectives in chemistry education built upon conceptions from psychology (cognitive science). One such conceptualization that has shaped how we think about teaching in chemistry education was suggested by Johnstone and a group of over 80 scientists and science educators (Johnstone, 2006). Following over 30 years of a blend of action-oriented research and practice- based experiences, the group derived at what they termed as an "information processing model" to account for the process of learning in science, and in particular chemistry. Their model was underpinned by two conceptual understandings from cognitive science. First was the understanding that an individual's brain system comprises a working space, referred to as "working memory" (Greene & Hicks, 1984; Miller, 1994), where information is processed, where we do all our planning prior to actions, such as developing a problem-solving strategy, among other things. Accordingly, this "working memory" has a very limited capacity to accommodate and work with information (Anmarkrud et al., 2019; Mathy & Feldman, 2012; Miller, 1956, 1994; Sweller & Chandler, 1994). Second, the group also adopted and aligned their conceived model with a Gestalt theoretical perspective (Koffka, 1935, 2014), as well as ideas by Ausubel et al. (1978) about the importance of pre-existing knowledge.

The information processing model (shown in Fig. 2.5), a version by Johnstone (1993) comprises three parts (we can call them compartments), the perception filter, the working memory space (conscious processing part), and the long-term memory. Accordingly, what is presented to the learners (external stimuli), that is, the subject matter concepts, or other learning experiences, are perceived and filtered before being forwarded for processing in the working memory space. During the filtering process, "the learner attends to what is familiar, stimulating, interesting, surprising or exciting" (Johnstone, 2006, p. 55), a process that Johnstone's group described as largely dependent on what the learner already has held in the long-term memory. Once in the working memory space, information is stored temporarily, while being operated upon to make sense of it, to enable the learner to respond accordingly or otherwise store the information in the long-term memory. Moreover, according to the group, "sense-making operation" involves recalling information already stored in the long-term memory to interact with new incoming information (Johnstone, 2006, p. 56). The author (Johnstone's group) described three scenarios that could happen during the sense making operation, which implicate the direction teaching and learning of chemistry could consider taking. That is, the interaction between new and old information can result in correct associations (acceptable understandings of the intended scientific concepts), or fault/incorrect associations, or a situation when there is no old information to relate to from the long-term memory and hence no sense making is made possible. For this third scenario, the group concluded that if the learner assesses new information as important, storage in the long-term memory proceeds but in form of rote learning.

Johnstone and his group's conceptions above which largely represent cognitivists thinking about learning (Chinn & Iordanou, 2023), create a thinking framework for us to start to understand and thereby confront the question of "*how students' (should) learn" in the context of chemistry teaching*. Let us start with the idea of pre-existing knowledge (Ausubel et al., 1978; Koffka, 1935) and that of a working memory with limited storage and processing capacity (Anmarkrud et al., 2019; Mathy & Feldman, 2012; Miller, 1994; Sweller & Chandler, 1994). First, Johnstone and his group concluded that without prior experiences to relate

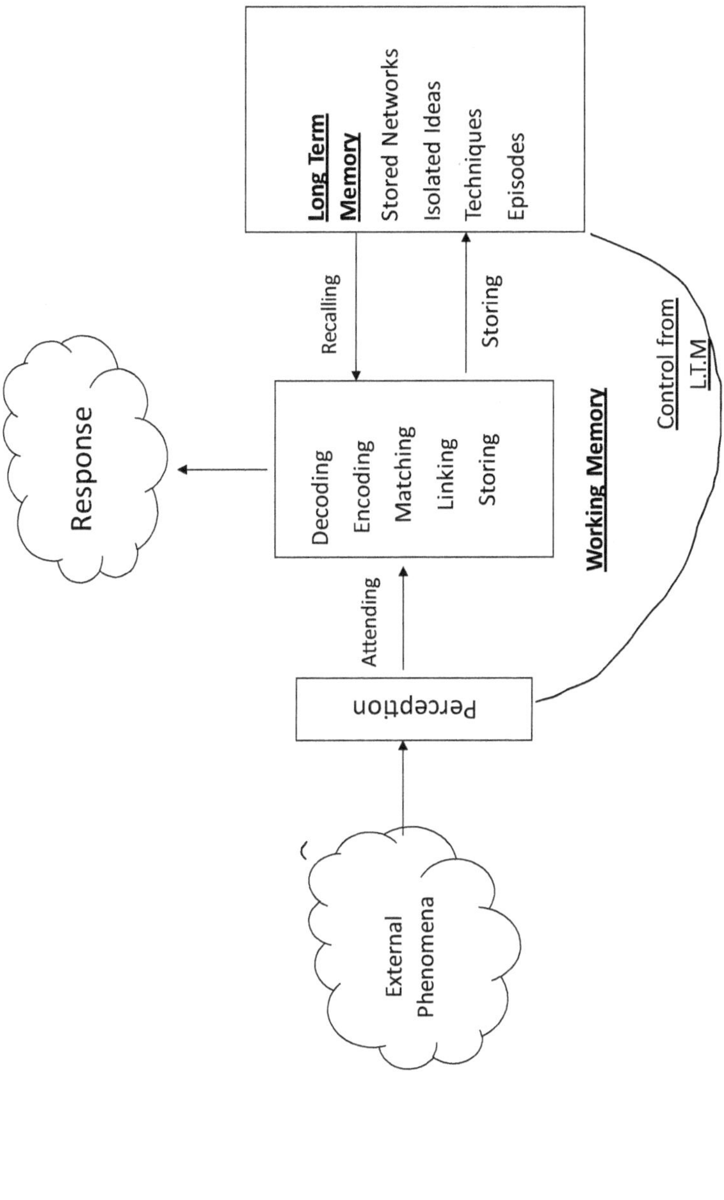

Fig. 2.5 Johnstone's version of Information processing reproduced with permission from Johnstone (1993). Copyright (1993), American Chemical Society

to from the long-term memory, new information will not find an anchoring point and sense making would be difficult. The student will thus need to find alternative ways to deal with new information, of which, one option is to discard or at best to store it in the form of rote learning. This understanding brings to the fore not only the importance of prior experiences (pre-existing knowledge), but also the need to enable learners to visualize, appreciate, and understand the value or the goal of what is being learned. Pre-existing knowledge does not only imply students having background understandings of previous chemistry conceptions to be related to new concepts to allow progressive and coherent learning, but students could also find anchoring points for new knowledge from experiences that relate to the applicability of the new knowledge being introduced. This idea fits well with Ausubel's (2000) meaningful learning conception, that is, the need for a teacher to make visible to the students the value of the concepts being introduced/taught.

Moreover, Johnstone and his group expressed the possibility of having two categories of students in the same chemistry classroom, (i) the field dependent students—students who are easily affected by the field against which the learning is done, easily distracted by irrelevant material and having difficulty in discriminating between the "signal and the noise", and (ii) the field independent students' category, students who are independent of the field, "can sharply focus on the relevant and ignore the irrelevant noise" (p. 54). If we are to consider both categories when teaching, then we get to understand that we even have to go deeper in our understanding of what meaningful learning should entail. Taber (2013), in response to Johnstone's proposition to desist from teaching at the middle of the triangle, argued that the nature of chemistry itself may not provide for this luxury, but rather teachers should assess students' learning needs continuously and devise corresponding measures, including slowing the pace at which new concepts are introduced and providing opportunities to reinforce key ideas.

Taber's suggestion above also seems an appropriate response to the second aspect concerning the working memory space and its limitation. Johnstone and his group concluded that, ".... if there is too much information to hold, there is not enough space for processing and the system overloads and seizes up. Similarly, if complex processing is needed, little information can be held for processing. This may also lead to overload and unsuccessful processing and storage." (p. 56). In view of the above, it is vital for us to acknowledge the distinction between ourselves as chemistry experts, and our students as novices, and thus understand that what is trivial for us is not trivial for the learners. Unlike novices, expert chemists (teachers in this context), have accumulated a rich base of experiences in their long-term memory to refer to when confronted with a new but related chemistry problem. This stored body of experiences serves as a reference framework and enables experts to work through, process and arrive at new sets of knowledge quickly. Taber (2013) gives a simple example of a neutralization reaction involving a strong acid (H_2SO_4) and strong base (NaOH) and illustrates how obvious and quickly the expert (teacher) can fit this into his/her familiar schema

(p. 163). Moreover, even with a mere mention of the statement "acid–base neutralization reactions", the teacher has already an organized structural (conceptual) map developed for this, unlike the student.

In Johnstone's (2006) report, particularly focusing on the works of El-Banna and Johnstone (1986), we see the action of "chunking" being invoked by some students as a way to work around overloading issues that tend to happen in the working memory space. From El-banna's study, students, either after being taught or without any training, managed to come up with strategies for breaking "over-size" problems into smaller sub-tasks (or chunks)" and sequenced these in a way that resulted in reaching the desired solution (p. 60). Johnstone and his group argued that students can be taught to devise strategies to chunk information, as long as the teacher holds the understanding that chunking strategies in most cases promote algorithmic performances more as opposed to promoting descriptive performances.

As briefly mentioned earlier, Johnstone and his group's conceptions as elucidated above are understandings about learning that largely draw on cognitivists views (see for example; Chinn & Iordanou, 2023). The focus is largely about the individuals' cognitive investment in terms of getting to learn chemistry, that is, drawing on existing experiences to acquire and store mental representations (concepts, models and schemas). However, the process that results in learning is not solely the work of an individual's cognitive investment but also involves the cultural and social investment. This where participationist views about learning become important (Danish & Gresalfi, 2018). Accordingly, students need to participate in social and cultural activities and learning then occurs through enculturation, discourse and identity developments among interacting members. Vygotsky's sociocultural theory of learning (Leach & Scott, 2003; Vygotsky, 1978) is one prominent theory in science education that emphasizes knowledge construction through social interaction and participation in cultural practices. We cannot ignore the role of social and cultural contexts in teaching and learning chemistry. Instead, we need to draw on both perspectives to be able to plan effective teaching. Students will need opportunities for individual meaning-making framed within the cognitivists' understandings of the limited capacity of the memory space, as well as opportunities for social interactions that will result in developing desired understandings.

2.2.1 What Do These Understandings About How Students Learn in the Context of Chemistry Mean to Us? (Teachers, Trainees and Educators)

During your teacher training, you will encounter huge amounts of educational literature with numerous suggestions about how to plan and teach effectively. You will also realize after a while that there is always that gap between recommendations about teaching and what happens in real teaching situations. The success of your teaching will largely be influenced by your personal attributes, such as intuition, emotional control and stability, efficacy, reflection ability, innovativeness, willingness to learn, re-learn, and adopt new strategies, among other key skills. While these attributes will be key to the success of your teaching, they are rendered inactive without a knowledge base that provides the knowledge, tools and resources needed to operationalize the teaching process. The teacher's knowledge about how students learn is a vital knowledge base that will provide you with a variety of thinking directions to take before, during and after your teaching.

We have seen that knowledge about how students learn not only comprises learning theories, but it is tightly linked with the nature of the subject itself. In other words, the theoretical learning views must be tailored to how the nature of chemistry is conceptualized. You will therefore be expected to make adaptations in the form of transforming general views about learning to suit your understanding of the nature of chemistry. Some chemistry educators like Cooper and Stowe (2018); Taber (2013), who have made a synthesis of the relationship between the nature of chemistry and students' learning, have come up with suggestions such as, the need to limit the information teachers present to students at any one time, to support students to relate new teaching to already existing understandings (experiences), plan to reinforce new learning over a given period to allow enough time building and strengthening connections between new and existing information and use scaffolding to support students' gradual learning of desired concepts, among other things (p. 166). As earlier mentioned, these and other suggestions that you will find in literature are not written permanent rules or procedures to be followed. You will time and again find yourself having to make necessary adjustments in your teaching, and hence the importance of your personal attributes (skills and abilities). Moreover, what you will need in principle is a rich knowledge base, a source of information (theoretical perspectives) about students learning, and about the nature of chemistry. Once you have background knowledge (perspectives and ideas) about how students (learners) come to understand chemistry principles, given the nature of chemistry, and what could be the possible hindrances to this knowledge construction, it will become easier for you to ignite and use your personal attributes as mentioned above to achieve the level of teaching practice you desire.

References

Agarkar, S., & Brock, R. (2017). Learning theories in science education. In K. S. Taber & B. Akpan (Eds.), *Science education: An international course companion* (pp. 93–103). SensePublishers. https://doi.org/10.1007/978-94-6300-749-8_7

Anmarkrud, Ø., Anette, A., & Bråten, I. (2019). Cognitive load and working memory in multimedia learning: Conceptual and measurement issues. *Educational Psychologist, 54*(2), 61–83. https://doi.org/10.1080/00461520.2018.1554484

Ausubel, D. P. (2000). The nature of meaning and meaningful learning. In D. P. Ausubel (Ed.), *The acquisition and retention of knowledge: A cognitive view* (pp. 67–100). Springer Netherlands. https://doi.org/10.1007/978-94-015-9454-7_4

Ausubel, D. P., Novak, J. D., & Hanesian, H. (1978). *Educational psychology : A cognitive view* (2d ed.). Holt.

Barnett, J., & Hodson, D. (2001). Pedagogical context knowledge: Toward a fuller understanding of what good science teachers know. *Science Education, 85*(4), 426–453. https://doi.org/10.1002/sce.1017

Brown, S. A., & McIntyre, D. (1993). *Making sense of teaching*. Open University.

Bucat, R. (2004). Pedagogical content knowledge as a way forward: Applied research in chemistry education [10.1039/B4RP90025A]. *Chemistry Education Research and Practice, 5*(3), 215–228. https://doi.org/10.1039/B4RP90025A

Bucat, R. (2005). Implications of chemistry education research for teaching practice: Pedagogical content knowledge as a way forward. *Chemistry Education International, 6*(1), 1–2.

Carter, K. (1990). Teachers' knowledge and learning to teach. In W. R. Houston (Ed.), *Handbook of research on teacher education* (pp. 291–310). Macmillan.

Chinn, C. A., & Iordanou, K. (2023). Theories of learning. In N. G. Lederman, D. L. Zeidler, & J. S. Lederma (Eds.), *Handbook of research on science education* (pp. 89–120). Routledge Taylor & Francis Group.

Connelly, M. F., & Clandinin, J. D. (1985). Personal practical knowledge and the modes of knowing: Relevance for teaching and learning. In E. Eisner (Ed.), *Learning and teaching the ways of knowing* (Vol. 84, pp. 174–198). University of Chicago Press.

Cooper, M. M., & Stowe, R. L. (2018). Chemistry education research—From personal empiricism to evidence, theory, and informed practice. *Chemical Reviews, 118*(12), 6053–6087. https://doi.org/10.1021/acs.chemrev.8b00020

Danish, J. A., & Gresalfi, M. (2018). Cognitive and sociocultural perspectives on learning: Tensions and synergy in the learning sciences. In F. Fischer, C. E. Hmelo-Silver, S. R. Goldman, & P. Reimann (Eds.), *International handbook of the learning sciences* (pp. 34–43). Routledge.

El-Banna, H., & Johnstone, A. H. (1986). Capacities, demands and processes–a predictive model for science education. *Education in chemistry*, (23), 80–84.

Erduran, S., & Kaya, E. (2019). *Transforming teacher education through the epistemic core of chemistry*. Springer.

Ertmer, P. A., & Newby, T. J. (2013). Behaviorism, cognitivism, constructivism: Comparing critical features from an instructional design perspective. *Performance Improvement Quarterly, 26*(2), 43–71. https://doi.org/10.1002/piq.21143

Geddis, A. N. (1993). Transforming subject-matter knowledge: The role of pedagogical content knowledge in learning to reflect on teaching. *International Journal of Science Education, 15*(6), 673–683. https://doi.org/10.1080/0950069930150605

Greene, J., & Hicks, C. (1984). *Basic cognitive processes*. Open University Press.

Gundem, B. B. (2004). Understanding european didactics. In B. Moon, M. Ben-Peretz, & S. A. Brown (Eds.), *Routledge international companion to education* (1st edition. ed.). Routledge.

Hamilton, D. (1999). The pedagogic paradox (or why no didactics in England?). *Pedagogy, Culture & Society, 7*(1), 135–152. https://doi.org/10.1080/14681369900200048

Johnstone, A. H. (1982). Macro and microchemistry. *School Science Review, 64*(227), 377–379.

Johnstone, A. H. (1991). Why is science difficult to learn? Things are seldom what they seem. *Journal of Computer Assisted Learning, 7*(2), 75–83. https://doi.org/10.1111/j.1365-2729.1991.tb00230.x

Johnstone, A. H. (1993). The development of chemistry teaching: A changing response to changing demand. *Journal of Chemical Education, 70*(9), 701. https://doi.org/10.1021/ed070p701

Johnstone, A. H. (2006). Chemical education research in Glasgow in perspective [10.1039/B5RP90021B]. *Chemistry Education Research and Practice*, 7(2), 49–63. https://doi.org/10.1039/B5RP90021B

Kansanen, P. (2009). Subject-matter didactics as a central knowledge base for teachers, or should it be called pedagogical content knowledge? *Pedagogy, Culture & Society, 17*(1), 29–39. https://doi.org/10.1080/14681360902742845

Kayima, F. (2022). The role of reflective practice in mediating development of pre-service science teachers' professional and classroom knowledge. *Interdisciplinary Journal of Environmental and Science Education, 18*(1), e2262. https://doi.org/10.21601/ijese/11364

Koffka, K. (1935). *Principles of gestalt psychology*. Routledge & Kegan Paul.

Koffka, K. (2014). *Principles of gestalt psychology*. Mimesis International.

Leach, J., & Scott, P. (2003). Individual and sociocultural views of learning in science education. *Science & Education, 12*(1), 91–113. https://doi.org/10.1023/A:1022665519862

Lederman, N. G., & Lederman, J. S. (2015). The status of preservice science teacher education: A global perspective. *Journal of Science Teacher Education, 26*(1), 1–6. https://doi.org/10.1007/s10972-015-9422-7

Loughran, J. J. (2002). Effective reflective practice: In search of meaning in learning about teaching. *Journal of Teacher Education, 53*(1), 33–43. https://doi.org/10.1177/0022487102053001004

Mathy, F., & Feldman, J. (2012). What's magic about magic numbers? Chunking and data compression in short-term memory. *Cognition, 122*(3), 346–362. https://doi.org/10.1016/j.cognition.2011.11.003

Miller, G. A. (1956). The magical number seven plus or minus two: Some limits on our capacity for processing information. *Psychological Review, 63*(2), 81–97.

Miller, G. A. (1994). The magical number seven, plus or minus two: Some limits on our capacity for processing information. *Psychological Review, 101*(2), 343–352. https://doi.org/10.1037//0033-295x.101.2.343

Mishra, P., & Koehler, M. J. (2006). Technological pedagogical content knowledge: A framework for teacher knowledge. *Teachers College Record, 108*(6), 1017–1054. https://doi.org/10.1111/j.1467-9620.2006.00684.x

OECD. (2017). *Pedagogical knowledge and the changing nature of the teaching profession*. https://doi.org/10.1787/9789264270695-en

Schubert, W. H., & Ayers, W. (1992). *Teacher lore: Learning from our own experience*. Longman.

Shulman, L. S. (1986). Those who understand: Knowledge growth in teaching. *Educational Researcher, 15*(2), 4–14.

Shulman, L. S. (1987). Knowledge and teaching: Foundations of the new reform. *Harvard Educational Review, 57*(1), 1–23. https://doi.org/10.17763/haer.57.1.j463w79r56455411

Simon, B. (2013). Why no pedagogy in England. In A. S. Mayes & B. Moon (Eds.), *Teaching and learning in the secondary school*. Routledge.

Sweller, J., & Chandler, P. (1994). *Why some material is difficult to learn* [185–233]. Hillsdale.

Taber, K. S. (2013). Revisiting the chemistry triplet: drawing upon the nature of chemical knowledge and the psychology of learning to inform chemistry education [10.1039/C3RP00012E]. *Chemistry Education Research and Practice, 14*(2), 156–168. https://doi.org/10.1039/C3RP00012E

Tro, N. J. (2017). *Chemistry: A molecular approach* (4th ed. (Global ed.). ed.). Pearson.

Ulferts, H. (2019). *The relevance of general pedagogical knowledge for successful teaching*. https://doi.org/10.1787/ede8feb6-en

van Driel, J. H., Verloop, N., & de Vos, W. (1998). Developing science teachers' pedagogical content knowledge. *Journal of Research in Science Teaching, 35*(6), 673–695. https://doi.org/10.1002/(sici)1098-2736(199808)35:6%3c673::Aid-tea5%3e3.0.Co;2-j

Vygotsky, L. S. (1978). *Mind in society: Development of higher psychological processes*. Harvard university press.

Preparing to Teach Chemistry 3

Abstract

In this chapter, the teacher's framework and governing tools (resources) for the teaching process are addressed. It is with these tools and resources, namely, the curriculum, the lesson plan, the instructional methods, teaching resources and aids, that a teacher can develop and implement teaching as desired or intended. The chapter will also discuss the different aspects and dynamics involved in the process of planning for teaching and managing the teaching process, as well as the relationship between teacher and students, their place and role in the teaching–learning process. Elements central to managing the teaching process such as motivation, interest, teacher abilities and attributes are also taken up and discussed in the chapter. Overall, the chapter seeks to equip the chemistry teacher (trainee) with a knowledge base for effective planning, implementation and management of the teaching process within a pre-planned working framework.

3.1 The Curriculum as a Teaching Guiding Framework

3.1.1 Defining What a Curriculum Is

The term curriculum and its use in educational contexts is not new to us (the teachers, educators or trainees), and whereas we find in educational literature varied conceptualizations of what the curriculum is and what it comprises, we will in this section stick to the common and familiar use of this word as we have come to understand it throughout our schooling years. Curriculum is often defined as a document comprising and describing all the planned learning experiences of a school or educational institution, that is, the planned goals, objectives, the content, the learning activities and evaluation procedures for a given course, study program or educational level (Bobbitt, 1918). It is a written document (in print or digital

F. Kayima, *Preparing for Chemistry Teaching*, Springer Texts in Education,
https://doi.org/10.1007/978-3-031-97122-8_3

format) comprising the contents of a subject, concepts and tasks to be acquired, planned activities, the desired learning outcomes and experiences, all of which are informed or underpinned by a society's agenda and its considered values (Klein, 1991). Some educators generally describe the curriculum as a written plan outlining what students will be taught (a course of study), and refer to all the courses offered in a school in a particular area of study (Glatthorn et al., 2018; McBrien & Brandt, 1997).

Goodlad and his colleagues (1979) provide a widely embraced conceptualization of the curriculum into five transition phases or forms, *the ideal*, *the formal*, *the perceived*, *the operationalized*, and *the experienced curriculum*. *The ideal curriculum* represents what the developers' intentions are (society's interest) and constitutes that which the developers desire, know, and believe ought to be taught to future generations. *The ideal* becomes *the formal*, when transformed into a form that is accessible to stakeholders in learning institutions, e.g., in the form of textbooks and or different kinds of instructional materials (e.g., stories, images or other relevant activities). *The perceived curriculum* reflects the teacher's interpretive frame (perception) of what ought to be taught and learned by students, while the *operational curriculum* is what is taught or delivered by the teacher. Finally, the *experienced curriculum* represents what is learned by the students. Acedo and Hughes (2014) describe four learning facets of a curriculum leaving out Goodlad's *experienced curriculum facet*, that is, "*the intended curriculu*m (what we intend students to learn), *the written curriculum* (how these intentions are laid out*), the taught curriculum* (what happens in the classroom) and *the hidden curriculum* (subconscious, institutional intentionality)" (p. 504). As you will find in several educational sources (e.g.; in general educational textbooks and published articles), some educators conceptualize the curriculum in its simplest form and merge *the ideal* and *the formal* into *the planned curriculum* (what developers desire or intend), *the perceived* and *operational* into *the delivered or implemented curriculum* (what is organized by school administrators and taught by teachers), and then *the experienced curriculum* (what is learned by students). Moreover, different countries adopt different structures and models for their curricula documents. Some countries have clear distinctions between *the formal and the operationalized* curricula. The latter is normally referred to as the school curriculum comprising all learning experiences a given school intends that its students should acquire (develop or obtain) and it reflects how the individual schools interpret the formal curriculum, extract and transform information into consumable forms by their teachers and other stake holders. From the school curriculum then the individual schools develop what is termed as *the school teaching syllabus*, a document detailing what topics or contents are to be covered in a particular subject. Moreover, some countries also have an examination syllabus as a separate entity detailing what topics or contents in a particular subject will be examined at the end of a course program or academic level.

Not all countries consider the above breakdowns of the national curriculum (formal curriculum). There are variations in how different countries structurally organize and break down their curricula from the formal level and down to the implementation phase. What is important and the focus here is how do we as teachers extract information from the formal curriculum, plan and execute a teaching that results in the achievement of intended teaching aims.

At the beginning of this section, we defined a curriculum as comprising at least five elements, the goals and objectives, the content, teaching methods or processes, the materials (teaching and learning resources and activities), and assessment methods. However, this is an open-ended description, and as you will find, different curricula from different countries vary in what elements are included. In the overview (Fig. 3.1) provided by the TIMSS 2015 Encyclopedia: Education Policy and Curriculum in Mathematics and Science (Mullis et al., 2016), we can clearly see this variation from one country to another. The overview shows some of the above mentioned elements in the science and mathematics curricula of different countries. As we see from the shown Fig. 3.1, four elements of curriculum are considered, goals and objectives, instructional methods or processes, materials, and assessment methods or activities. It is further visible that not all countries include all the elements in their national curricula documents. For example, the figure shows that countries like Norway, England, United states, among others, only state goals and objectives but do not give or dictate on what instructional methods or processes or what materials and assessment strategies would be used on what topics (note that this information is from a 2016 report and changes might have been made respectively within the last 9 years). For the other countries like Jordan and Turkey, etc., all the four elements are included in their individual curricula documents. This, as shown in Fig. 3.1, reflects the variations in teachers' expectations when it comes to extracting and using information from curricula documents.

3.1.2 The Chemistry Curriculum: Rationale and Orientations

Teachers being curriculum implementors, are expected to study and gain a full understanding of the rationales that lie behind the framing of any chemistry curriculum that is to be operationalized. This understanding forms a strong basis for effectively interpreting the curriculum by aligning the teacher's understanding and perceptions about teaching with the orientation or framework based upon which the curriculum is built. Mostly, developers of the chemistry curriculum have their focus on the question about what contributions chemistry education brings to society. It is often envisioned that a chemistry curriculum will (i) result into many young people progressing as chemical scientists, (ii) provide skills and knowledge usable in other areas other than a career in chemistry, and (iii) ensure and contribute to scientifically literate persons, who are capable of engaging in informed debates on issues that affect them at both a personal and societal level (for example promoting sustainable living- see Chapter 10). From science education research reviews, we find different possible models or orientations based upon which different chemistry curricula are framed. Curricula orientations often reflect different understandings, aspects or areas of concern which curriculum developers choose

TIMSS 2015 Mathematics & Science 8th Grade

Exhibit 8: **Components Prescribed by Mathematics and Science Curricula at the Eighth Grade**

Reported by National Research Coordinators

Country	Goals and Objectives		Instructional Methods or Processes		Materials		Assessment Methods or Activities	
	Mathematics	Science	Mathematics	Science	Mathematics	Science	Mathematics	Science
Armenia	●	●	●	●	●	●	●	●
Australia	●	●	○	○	○	○	○	○
Bahrain	●	●	●	●	●	●	●	●
Botswana	●	●	●	●	●	●	●	●
Canada	●	●	●	○	○	○	○	○
Chile	●	●	○	○	○	○	○	○
Chinese Taipei	●	●	●	●	●	●	●	●
Egypt	–	–	–	–	–	–	–	–
England	●	●	○	○	○	○	○	○
Georgia	●	●	●	●	●	●	●	●
Hong Kong SAR	●	●	○	●	○	●	○	●
Hungary	●	●	○	○	○	○	○	○
Iran, Islamic Rep. of	●	●	●	●	●	●	●	●
Ireland	●	●	○	●	○	○	●	●
Israel	●	●	●	●	○	○	○	○
Italy	●	●	●	●	○	○	○	○
Japan	●	●	●	●	●	●	○	○
Jordan	●	●	●	●	●	●	●	●
Kazakhstan	●	●	●	●	●	●	●	●
Korea, Rep. of	●	●	●	●	●	●	●	●
Kuwait	●	●	●	○	●	●	●	●
Lebanon	●	●	●	●	●	●	●	●
Lithuania	●	●	●	●	○	○	●	●
Malaysia	●	●	●	●	●	●	●	●
Malta	●	●	●	●	●	○	●	●
Morocco	●	●	●	●	●	●	●	●
New Zealand	●	●	○	○	○	○	○	○
Norway	●	●	○	○	○	○	○	○
Oman	●	●	●	●	●	●	●	●
Qatar	●	○	○	○	●	●	●	○
Russian Federation	●	●	○	○	○	○	○	○
Saudi Arabia	●	●	●	●	●	●	●	●
Singapore	●	●	●	●	●	●	●	○
Slovenia	●	●	●	●	●	●	●	●
South Africa	●	●	●	●	○	○	●	●
Sweden	●	●	○	○	○	○	○	○
Thailand	●	●	●	●	●	●	●	●
Turkey	●	●	●	●	●	●	●	●
United Arab Emirates	●	●	●	●	●	●	●	●
United States	●	●	○	○	○	○	○	○
Benchmarking Participants—Responses Pertain to Benchmarking Provinces/Emirates/States								
Buenos Aires, Argentina	●	●	●	○	○	○	●	○
Ontario, Canada	●	●	●	○	○	○	●	●
Quebec, Canada	●	●	●	●	○	○	●	●
Abu Dhabi, UAE	●	●	●	●	●	●	●	●
Dubai, UAE	●	●	●	●	●	●	●	●
Florida, US	●	●	○	○	○	○	○	○

● Yes
○ No

A dash (–) indicates data not provided.

SOURCE: IEA's Trends in International Mathematics and Science Study – TIMSS 2015

Fig. 3.1 Overview of the science and mathematics curricula elements of different countries as provided in The TIMSS 2015 Encyclopedia: Education Policy and Curriculum in Mathematics and Science, Reproduced with permission from IEA—The international Association for Evaluation of Educational Achievement

to consider when framing their curriculum. These orientations not only inform or guide the framing of the curriculum but can also guide teachers when operationalizing the contents of the curriculum. Among the examples of chemistry curricula orientations include, (i) a chemistry curriculum oriented towards the structure of the discipline—that is, a curriculum that mirrors the contents of a typical chemistry textbook, (ii) a curriculum oriented towards context-based teaching conceptualizations, or (iii) a chemistry curriculum driven by environmental issues, history of science, or a technology and innovation driven chemistry curriculum (Eilks et al., 2013).

About a few years ago (by July 2025), the Royal Society of Chemistry (RSC) proposed, what was called a framework for "a successful chemistry curriculum", premised on the group's understanding of the contributions of chemistry education to society as well as how the group understands and defines chemistry as a discipline (Gibney, 2018). Starting with the common understanding that chemistry is concerned with studying the composition and properties of matter and how and why it undergoes change, RSC submits that the body of chemistry knowledge has been built up through centuries of observation, experiment and reasoning, and that our understanding of matter is still ongoing through empirical study and ways of reasoning. The RSC group adds that chemical scientists build on the existing knowledge body, create new materials, and derive at new findings that help to understand and improve the world. Premised on this understanding, the RSC group proposes a chemistry curriculum framework that comprises three chemistry areas, (i) *Chemistry as a science*—the practices of thought and enquiry that we use to develop new understandings, (ii) *Chemical concepts*—the body of knowledge about matter and properties that the practice of chemistry has developed over time, and (iii) *Chemistry and the world*—the way chemistry is used to understand and affect the world in which we live (Gibney, 2018, pp. 30–31). Indeed, this curriculum framework by RSC covers the understandings presented in Chapter 1 regarding the three knowledge dimensions of a full body of chemistry, encompassing both the ontological chemical knowledge and the epistemic aspects of chemistry. Considering the RSC's curriculum development work, we visualize the interplay between curricula rationales and curricula orientations, and the importance of understanding the discipline of chemistry in terms of not only its nature, but also its place and contribution to society.

3.1.3 From the Formal to the *Delivered or Operationalized* Curriculum

When you study the chemistry curriculum document for your country, it is possible to identify the rationale behind its development and what kind of orientation or underlying conceptual framework that underpinned its development. For example, it is most likely that developers of the curriculum for your country have greatly advocated for an inquiry-based teaching orientation or altogether embracing a competency-based curricula orientation (Morcke et al., 2013; Soare, 2015). Even when this example does not necessarily reflect the ideas behind the development of the chemistry curriculum for your country, what is clear here is that the amount of work and energy you will invest in planning and extracting information from the formal curricula documents and transforming it into usable forms will depend on what kind of curriculum framework your country has developed or adopted. As it is shown in Fig. 3.1, chemistry (or science) curricula by different countries take different forms and structure in terms of what elements are considered. Some leave the teacher with the freedom to decide what teaching philosophy to adopt and employ, including selecting the teaching and assessment methods that will enable attainment of the target curriculum goals. In Norway for example, in addition to stating the competence aims, the current (2022) chemistry curriculum includes only what is considered core elements in chemistry at the high school level with brief but general information about what each core element is about, that is, *(i) practices and ways of thinking in chemistry, (ii) chemical bonds and structures, (iii) chemical reactions, (iv) applied chemistry*. It does not, however, specify or suggest methods or activities to be used to cover the intended competence aims and neither does the framework give a detailed account of the relevant chemistry topics to cover the intended competence aims. The chemistry teacher is thus expected to use his/her competence and guided by both the list of core elements and competence aims, to decide upon the relevant and suitable chemistry topics (contents) that will cover individual competence aims of the curriculum. Several other countries have a similar or related curriculum framework structure, where the curriculum provides an open room for the teacher to display his/her creativity in as far as deciding on what topics to teach and how they will be taught, as long as there is sufficient assurance that the intended goals of the curriculum will be attained.

On the other hand, other frameworks are structured in a very closed way that dictates how the implementors of such curricula are supposed to approach the teaching. They include not only what learning goals, aims or experiences are intended, but also details of what specific chemistry topics or content to be taught to cover specific aims, as well as what nature of teaching methods, teaching aids and resources, are to be used to teach specific chemistry topics, and how students' learning will be assessed. For better understanding, I want to invite you to pause for a moment and look at the formal national curriculum document for your country, specifically for chemistry, and briefly examine its structure, identify and discuss what elements are included. Please note that, in getting to understand these fundamental differences in the structure and framing of different chemistry curricula, the aim is not to assess how good or bad the curriculum structure is, but to enhance your ability to explore, identify, and understand the fundamental ideas behind the development and framing of the formal curriculum. This understanding is important to how we (as teachers) should perceive, interpret and operationalize the curriculum.

3.2 Planning and Structuring Chemistry Lessons and Activities

3.2.1 The Scheme of Work and Lesson Plan

As already mentioned, part of your expected role as a classroom chemistry teacher will be to study and interpret the formal chemistry curriculum, extract and transform its contents into forms consumable at the learning level. This is where we come to talk about both the *scheme of work* and the *lesson plan* as two important framework planning tools that are useful in the process of breaking down and transforming curricula contents into usable forms.

3.2.1.1 The "Scheme of Work"

A scheme of work (or a scheme of learning or a learning programme as some have called it) can be thought of as a physical representation (a written document for example) of a certain way you can structure the teaching and learning and the assessment of your subject in a logical and progressive manner over a certain period. Some educators have called *a scheme of work* a physical document, but it can be presented in different forms. Moreover, regardless of the format chosen or used, the scheme of work represents a breakdown of curricula contents that you the teacher yourself make when you split the contents of the curriculum into deliverable units or several sessions. This stage of your planning for teaching is crucial and enables you to align the contents of the curriculum, e.g., specific curriculum target goals with suitable topics to use and cover target curriculum goals, find out what teaching resources (including methods, materials and other learning aids) would support the teaching if not provided, as well as find the most suitable assessment methods for specific learning outcomes. It is possible to plan or draft a scheme of work for a whole programme or a full term or semester, or a full academic year (three terms in most secondary schools). Most probably, you will carry out this planning as a group with your colleagues teaching the same subject at the same class level, or as a unit. However, you should also be prepared to plan and draft a scheme of work as an individual and rise to the occasion under such circumstances where you are the only chemistry teacher available in that given school.

Depending on where you will be working, the practice of planning for teaching will vary from one context to the other. In some educational settings, drafting *a scheme of work* is a collective task for the school and quite often, the department will set a day off specifically for planning the teaching for the entire academic year or semester. In other schools or contexts, the administration (or head of department) normally invites classroom subject teachers to submit a scheme of work for the entire term (semester) well before the term commences. Regardless of what structural organisation or strategic plan is followed, you will and are expected to actively take part in transforming curricula contents into usable forms and this is mostly the central goal of planning at this stage. When drafting a scheme of work, what you choose to include or exclude is a question of what elements are important to enable you to plan and implement your teaching in a given specified period (weeks, months, term, semester or entire year). For example, we noted earlier that in certain countries, the chemistry curriculum only specifies what competence aims should be attained in the curriculum as well as what core themes the developers conceived as central and relevant. The curriculum does not state the relevant contents to cover the core themes, and neither does it state what methods, activities, aids or assessment methods to employ during the teaching process. With this format of the chemistry curriculum, the teacher is expected to work out all these elements when planning and drafting a scheme of work. Moreover, s/he is expected to determine not only the relevant topics, but also draft concrete and feasible learning objectives targeting the attainment of specific curriculum target aims (goals).

Fortunately, in some countries, authors of classroom chemistry textbooks endeavor to align different topics in the textbooks with specific curriculum target aims (goals), and in so doing, they make the teachers' planning easier. Again, I want to invite you at this point to pause, obtain a sample of a secondary chemistry textbook used in your local school, briefly examine the contents to establish if the authors have structured the different topics by considering the governing chemistry curriculum operationalized in your educational context. Find out if they have concretely specified and aligned the different chemistry topics with different competence goals as they appear in the local school chemistry curriculum.

Regardless of what nature of support is provided by local chemistry textbook authors in your country, when planning and drafting a scheme of work for a specified grade, program or class level, you will need to establish and decide upon the following.

i. What duration in terms of number of days, weeks, months, or term, do you intend to cover in your plan?
ii. What curriculum competence aims (goals) will be targeted within that specified period (duration)?
iii. What specific chemistry topics will be covered as you seek to address the curriculum competence aims?
iv. How many sessions of what duration do you intend to include for the specified period?
v. What are the available learning resources (relevant textbooks, lab facilities, digital tools, other available aids and resources) that could be used when implementing the individual planned sessions
vi. Based on the nature of the chosen topics as well as the available resources, what teaching approaches are possible to include in the plan?
vii. What are the potential assessment strategies based on the available learning resources and the nature of included topics?

Moreover, from your general education courses (e.g. general pedagogical courses), you might have been told that you also need to specify the learning outcomes or objectives, students' as well as teachers' activities when drafting a scheme of work.

You will however find that it is very superficial and not realistic to include these elements at this time of the planning phase, and that you would rather "keep these at the back of your minds". It will be very difficult for you to concretely define what or how your role as a teacher will unfold over a number of sessions that you are including in your scheme of work. Neither will it be realistic to concretely state what your students will be doing in each one of the sessions you are including in your scheme of work. These elements can properly and concretely be planned for when working out a lesson plan as we will discuss later. Thus, it is important to mainly focus on feasibly measurable elements such as those listed above.

3.2.1.2 The Lesson Plan

Once you have drafted your scheme of work (program for teaching) either individually or collectively as a unit, what follows is the daily or routine planning for the single sessions as included in your main framework plan (the scheme of work). The session plan (what I call lesson plan) is a working document that details what you intend to teach in a specified time period, why you intend to teach it (aims/objectives and learning outcomes), how you intend to teach it, with what tools, methods and resources, as well as how you intend to assess the attainment of the intended aims for teaching what you intend to teach. Figure 3.2[1] represents a summary of the different aspects to consider when planning for a teaching unit. Moreover, you could also include details about what your role will be and that of your students. Since this is the working document that guides you as you implement your teaching, you are free to include other personal elements, such as a personal evaluation space, where you capture your real time reflections about the unfolding of the lesson, or a brief statement describing the nature of your students, including both yours and their expectations. I noted earlier that it was not realistic when drafting a scheme of work to concretely specify learning objectives/outcomes or state what your (teacher) and students' role would be. It is during lesson planning that these elements can concretely be drafted and made feasible.

If we make a summary of the elements above, we come to the following guiding list.

i. What you intend to teach—content details—topic/subject
ii. Why you intend to teach it—specific aims/objectives, and learning outcomes
iii. How you intend to teach it—selected teaching approach/methods
iv. What tools and other resources—teaching resources (textbooks and other learning aids)
v. How you intend to assess the attainment of intended aims or learning outcomes—selected assessment method/approach: this also includes assessment for learning as discussed in Chapter 6

[1] Figure 3.2 is also used in Chapter 11 when discussing the different elements and expectations from school practice.

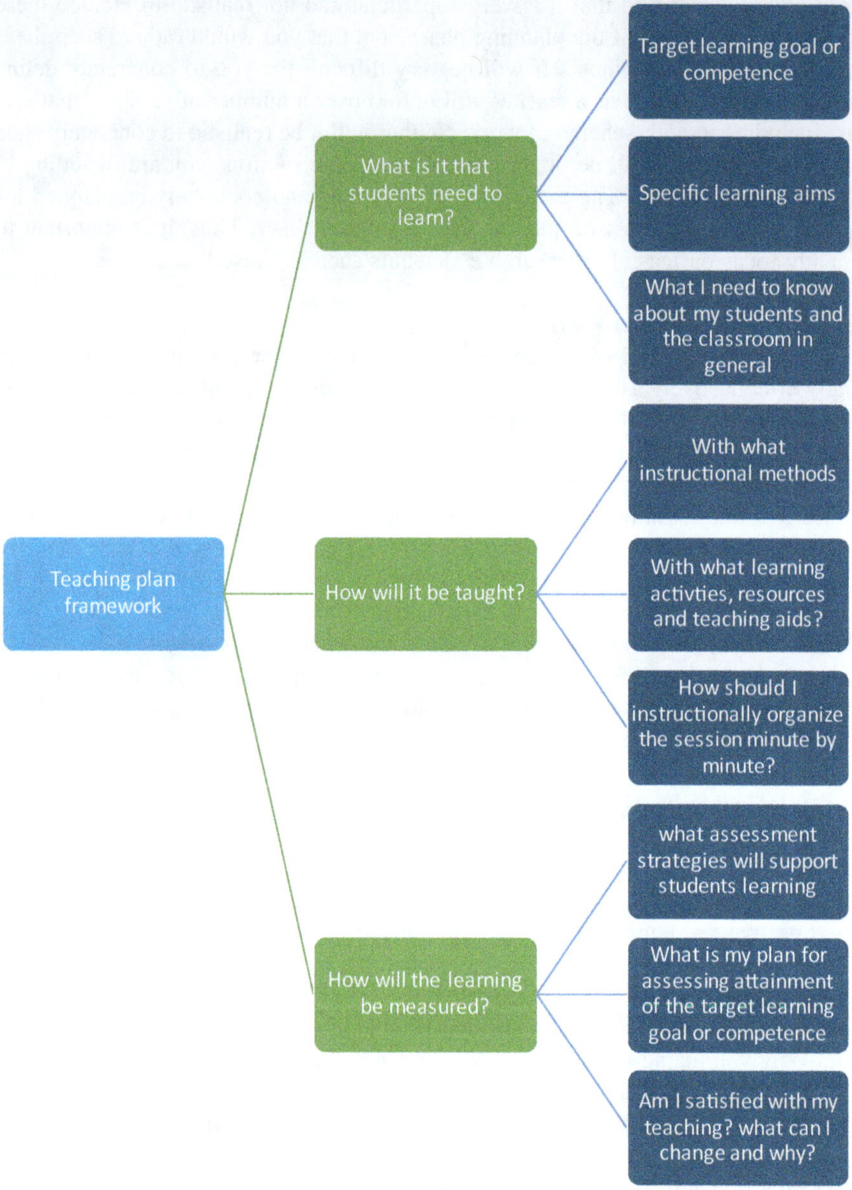

Fig. 3.2 A teaching plan framework comprising elements to be considered when you plan a lesson

vi. How do you intend to assess/evaluate the progress of own teaching—self-evaluation space
vii. How you describe your students' expectations or motivation to learn

As a recap of the aspects presented so far in this chapter, consider the following reflection questions.

a. Explore and discuss the practice in your country in terms of how schools structurally organize the teaching process. Focus on how the chemistry curriculum document is operationalized.
b. Discuss whether planning and developing a scheme of work is part of the practicing culture in your context? If not, how is the teacher's planning for teaching organized starting with curriculum and down to the implementation phase in the classroom?
c. Discuss the potential support both the scheme of work and lesson plan provide to teachers in their teaching? What challenges do you envision in being asked to prepare and have in place at least one of these two documents?

3.3 Instructional Approaches, Methods, and Strategies

When planning *what to teach* as a way to attain the target curriculum goals, we must answer the question of ***how we intend to or will teach whatever aspects or contents we have planned to teach.*** Thus, we need a teaching approach, a strategy or a method, with which we will navigate the teaching and learning process to reach our end goal. Today (in 2025), there is a myriad of educational resources providing a rapporteur of teaching approaches, methods, strategies, as well as several didactic models which are potentially useful and applicable to chemistry teaching. Quite often, the distinction between ***instructional approaches, methods and strategies*** is not made clear in educational literature. However, this should not be considered a problem that impedes us from finding the most suitable ways to teach and achieve our target aims. That said, knowledge of how different educators in both science and general education conceptualize the above three terms, as well as the examples given when using the terms, could be useful in informing our selection of the most appropriate ways to approach the teaching and learning of chemistry.

In some science education books and published articles, you will find the terms, *methods* and *strategies* being used interchangeably or close to one another. For example, in a chapter on "General instructional methods and strategies" by Treagust (2007) in the *Handbook of research on science education* by Lederman and Abell (2007) and the same chapter by Treagust and Tsui (2014) in a later version (edition) by Lederman and Abell (2014), the authors use the terms *instructional methods* and *strategies* together in the same sentence without pointing to any inherent differences in meaning between these two terms. In the 2007 edition in this chapter, the author (Treagust, 2007) writes that, *"instructional methods and strategies can be organized in terms of the amount of direct control that teachers and instructors have over their implementation (p. 373*), whereas in the same chapter in the 2014 edition, the authors write that, *"Science teachers use a multiplicity of instructional methods and strategies when teaching science classes .... **These strategies** range from primarily didactic or teacher-centered to those that are primarily student -centered or learner-centered. (p. 303).* In this latter statement, the authors appear to drop the term ***"instructional methods"*** and proceed only with the term ***"strategies",*** an indication that they consider no distinction between these two terms. However, in both editions, the authors describe different theoretical considerations understood to influence the different strategies and methods employed in science teaching. These different theoretical considerations constitute what other educators characterize as ***instructional approaches*** as I briefly express in the later paragraphs.

In another general education didactic textbook, *"Teaching strategies: A Guide to effective instruction"* by Orlich et al. (2012), the authors categorically state that they use the terms *teaching strategy, method, and procedure* as **synonyms**. Moreover, in an OECD report on *"teaching strategies for instructional quality"* (Noémie et al., 2016), the authors only use the term *"teaching strategies"*, and distinguish these into *active learning, cognitive activation*, and *teacher-directed instruction.* It is reasonable to conclude from the above that the terms, *instructional methods and strategies* are used as *umbrella* terms and at this conception level, their use focuses on the whole of the process undertaken to accomplish the target goal of the teaching.

As I mentioned earlier, as chemistry teachers, knowledge about the possible distinctions between the terms, *instructional approaches, methods and strategies* can be crucial for decision making during the planning phase of our teaching. For this reason, it is also of important interest to us when we consider what aspects (descriptions, and examples) are considered by those who have sought to distinguish between these terms. Again, here we find several educational resources where the authors distinguish between the terms, *instructional approaches, methods and strategies* (e.g.; Akdeniz, 2016; Lang & Evans, 2006; Nasiba et al., 2021; Saskatchewan-Education, 1991; Vural, 2016). Figure 3.3 is a summary of the plausible distinctions between *instructional approaches, strategies and methods* as suggested or described in selected educational resources. As you closely study Fig. 3.3, pay attention to the first element, that is the *"instructional approaches"* and how this has been conceptualized. Accordingly, all our planning for teaching, including what kind of strategies or methods we choose, are (or will be) strongly influenced by our inherent views (beliefs or ideas) about teaching and learning (knowledge about how learning takes place), or our knowledge about the most effective teaching orientations, theories or models. Thus, while a clear distinction between methods and strategies may be blurred, we must and should have a good understanding of and about our own beliefs about teaching or the kind of philosophical orientations that should inform how we should approach the teaching and learning process.

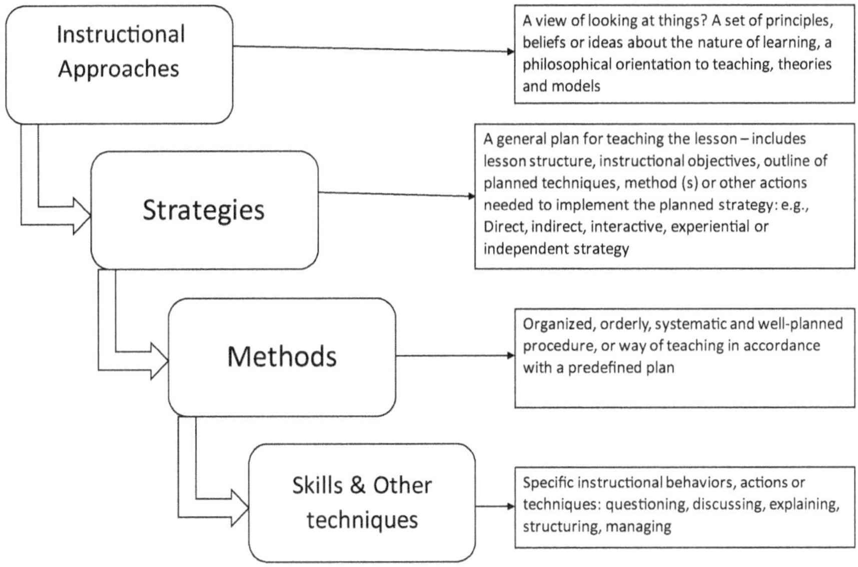

Fig. 3.3 Plausible distinctions between instructional approaches, strategies and methods: Ideas summarized from Akdeniz (2016); Lang and Evans (2006); Nasiba et al. (2021); Saskatchewan-Education (1991); Vural (2016)

As already noted (e.g.; Treagust, 2007; Treagust & Tsui, 2014), our choice of teaching strategies and methods lies along the continuum between teacher-centered and student-centered. This understanding (view) has not changed and is not about to change in the coming years even with such substantial developments in learning technologies that continue to shape the educational landscape. Indeed, there is a large body of educational literature presenting us with a repertoire of instructional methods and strategies to choose from depending on whether we adopt a student-centered or teacher-centered approach or both. Presently, in 2025, if asked to give your views about what teaching methods and strategies are associated with productive (effective or positive) learning effects (outcomes or benefits), you will most likely point to strategies or methods that actively involve learners in the teaching learning process. Moreover, most science education research reports within the last 4 decades including those published in chemistry education journals, strongly emphasize, some with empirical evidence, the use of learner-centered strategies or methods, such as inquiry-based learning, problem-based learning, project-based learning, flipped classrooms, discussions, case studies, demonstrations, to mention but a few.

Undoubtedly, active involvement of learners and the use of learner-centered strategies and methods could result in productive learning. However, as chemistry teachers, we must understand that there is no direct correlation between students' activation (active participation) and students' attainment of target learning outcomes. Moreover, you will experience in real classroom settings that the necessity of adopting a teacher-centered approach at times in your teaching will be inevitable when you want to achieve target learning aims. This is why your philosophy or set of principles and beliefs that inform your views about teaching, becomes an important factor in your planning for teaching.

In Fig. 3.4, I have imagined a kind of conceptual roadmap that could be drafted in our minds as teachers when we start to think about and thereby work towards deciding upon a certain instructional strategy or method with which to approach the teaching of a certain chemistry topic. First, is the rich knowledge body of theoretical perspectives about teaching that we have accumulated mainly from pedagogical courses and other encountered experiences, for example the behaviorists, cognitivists and participationist theories (Chinn & Iordanou, 2023; Leach & Scott, 2003; Sweller & Chandler, 1994) as briefly mentioned in Chapter 2. Moreover, to these different learning perspectives, we also have or are expected to hold a good understanding of chemistry, its nature, as well as the teachability and learnability of individual chemistry concepts (Bucat, 2004; Geddis, 1993). From this rich body of theoretical conceptions, along with our own beliefs, ideas and shared or attained experiences about teaching, a teaching philosophy is developed that which reflects how we conceive, interpret and understand the process of teaching. For those who are not able to forge or develop their own teaching philosophy, having this knowledge base enables them to select, embrace, or subscribe to one or a combination of theoretical ideas about teaching, and use this in planning and framing their teaching. Consequently, deciding what method or strategy to be used in teaching is hugely influenced by one's underlying understandings (theoretical teaching perspectives), beliefs and ideas about teaching. As Fig. 3.4 shows, a teacher who wants to draw on constructivists learning perspectives and considers social interactions to play a part in students' learning, will strive to use methods and strategies that provide contexts where students socially interact with the aim of creating their own learning either as individuals or in fellowship with others.

We should not however forget that the nature of chemistry concepts themselves, your personal qualities (attributes) as a teacher, as well as the individual target learning aims or objectives, have an influence on what methods and strategies are possible in a given teaching context (see for example, Yoder et al., 2021). Your students will be varied in their abilities, motivation and other personal needs. Knowledge of this also informs and influences the selection or choice of your teaching method and strategies for a given topic, nature and level of students. For example, chemistry education research has produced some empirical results that indicate that low achieving students perform better when methods and strategies used in teaching are preferred by these students (e.g., Kousa et al., 2018). This for example could be one of the aspects to consider when you have full awareness of the cognitive limitations and differences among your students.

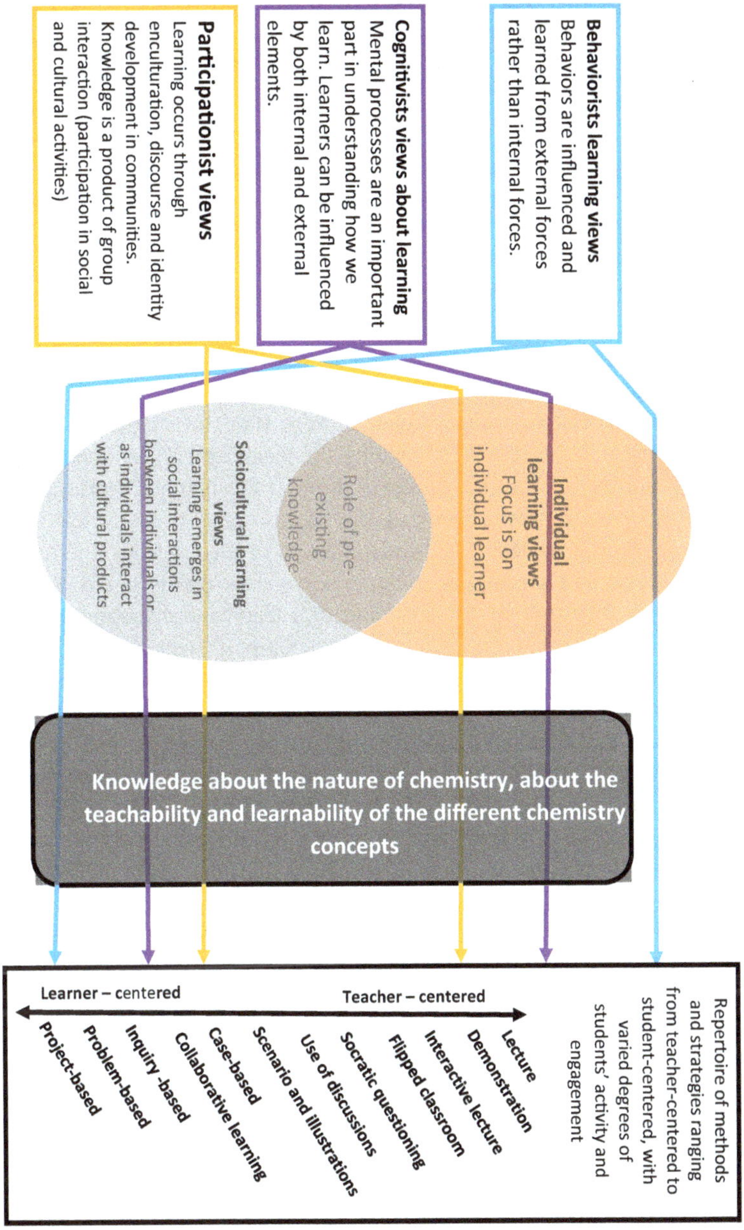

Fig. 3.4 Visualizing the role of theoretical perspectives (theories, beliefs, ideas or teaching orientation) when deciding teaching methods and strategies

3.4 Managing the Teaching Process

The classroom teacher is often likened to an artist, and teaching both an art and a science. Moreover, the classroom or any other physical space designated for teaching is often described and perceived as a theater where the artist, the teacher, does perform (Jay, 2005). Teaching is artistic in that there is no written or specific script to be followed by the teacher, while, it is a science in the sense that there are a number of research evidenced teaching practices (didactical and pedagogical knowledge and skills) that teachers can draw upon or employ and thereby potentially enhance students' learning (Marzano & Brown, 2009). Different teachers cannot be and are not expected to teach or act the same way during teaching, and neither are they expected to take up education research knowledge about teaching and use it in the same way. However, as artists, teachers are expected to inspire, transform and move students from their current state to a level where they are needed at the end of the lesson. Managing the teaching process therefore entails the teacher's ability to walk the students through the planned learning activities in a rather dynamic and ever-changing classroom context. The effective teacher productively reacts to students' real-time changes in their actions and behavior as the lesson unfolds and progresses, a dynamic process that is geared towards the attainment of the intended goals. Thus, a central feature to managing the teaching process concerns the teacher and student relationship, as well as a teacher's ability to facilitate a discourse where both teacher and students are continuously working together in varied and adaptive ways (e.g.; Ball & Cohen, 1999; Barnett & Hodson, 2001). Note that managing the teaching process is not the same as classroom management where a teacher is striving to maintain classroom order. Rather, the latter falls under the former and is simply one aspect of it.

Having planned what to teach and decided upon how it should be taught, you (the teacher) will then come on stage, to the place (theater) where the teaching happens. While on this stage, you will inevitably make real-time adjustments based on how students react in a dynamic classroom context. To navigate the complexities involved at this stage, it is crucial that you have a good understanding and or knowledge of and about the existing didactical and pedagogical practices, knowledge, skills and attributes that address teacher-students' relationships. This body of research grounded practices, knowledge, skills and attributes include among others, students' motivation and interest (e.g.; Bolte et al., 2013; Krapp & Prenzel, 2011), possible communicative approaches and feedback practices (e.g.; Chin, 2007; Mortimer & Scott, 2003), students' and teachers' use of questions (e.g.; Kawalkar & Vijapurkar, 2013; Kayima & Jakobsen, 2020; Kayima & Mkimbili, 2021), argumentation in science (e.g.; Duschl & Osborne, 2002; Ma et al., 2025; Osborne et al., 2004; Özdem Yilmaz et al., 2017), and other verbal or nonverbal communication tools. Note that, not only does this body of practices, knowledge, skills and attitudes feed into the management process of teaching, it is also crucial at the planning phase of the teaching. For example, when planning how to deal with and motivate students with low interest in science, when deciding upon the nature of questions to trigger students' thinking, etc. Figure 3.5 is a simple illustration of what I have called the "knowledge toolbox" for managing the teaching process to give you the teacher (trainee or educator) an overview of some of these elements as a basis for further exploration to expand your knowledge about these individual elements.

3.4.1 How is Managing the Teaching Process Practically Realized?

Well, it will be a far-fetched desire if we want to start to talk about having a commonly shared principle or procedure that teachers could follow when acting on stage. Students are different in terms of interests, strengths and limitations. This in turn implies every class is different and teachers work variously to suit such varied teaching situations. This is the very reason teaching is considered an art. Some educators have made attempts to understand what exactly happens during the action of teaching and how teachers should navigate the complexities involved in this teacher-student working relationship. Working independently and during different timelines, these educators (e.g.; Ball & Cohen, 1999; Barnett & Hodson, 2001; Kayima, 2022a; Schön, 1983, 1987) have described experienced teachers as being able draw on a different set of teacher knowledge comprising personal beliefs, past experiences, skills and attributes, to manage and navigate the teaching process. Schön (1987) described the uncertainty, the conflict, and the uniqueness of situations practitioners such as teachers must deal with during action situations, namely real-time teaching. Schön used the term "nonroutine" situations that occur during real-time actions and characterized them as partly indeterminate. Over three decades since Schön's conceptualizations, the experiences of managing the teaching process have not changed much, apart from having progressed in terms of having advances in research knowledge that feeds into teaching, better teaching resources, and technologies. The process of managing the teaching process accounts for the artistry of teaching (Harris, 1998) and demands that teachers "think on their feet" (Barnett & Hodson, 2001), constantly and variously adjust their positioning as the lesson progresses. This demand even places more emphasis on the need for teachers to examine their own personal attributes and skills, and even go further and seek to reinforce those attributes and skills that are learnable (Kayima, 2022b).

When we reflect about what happens in real-time teaching situations and looking at Fig. 3.5, a list of aspects (knowledge, skills, attributes, and practices) with the potential to support the teaching process, two questions arise, *(i) how should I as a chemistry teacher, deal with uncertainties of real-classroom situations?, (ii) how can I acquire or develop the knowledge, the skills and or practices listed in Fig. 3.5?.* For a reasonably acceptable response to the first question, we can go back in time, that is, about over 3 decades ago, and consider Schön's (1987) accounts that

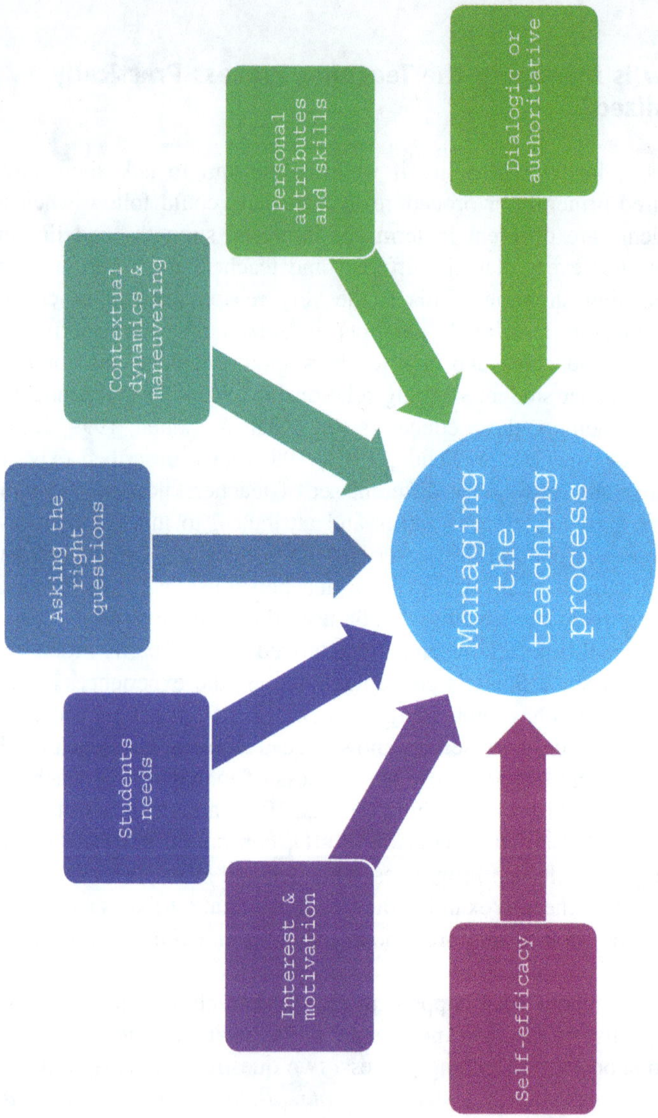

Fig. 3.5 Practices, knowledge, skills, attributes and practices that feed into managing the teaching process

point to a similar or related question. Schön expressed that when skillful practioners, teachers in our case, are faced with non-routine situations during real-time contexts, they.

> ....learn to conduct frame experiments in which they impose a kind of coherence on messy situations and thereby discover consquences and implications of their chosen frames. From time-time, their efforts to give order to situations provoke unexpected outcomes – "back talk" that gives the situation a new meaning. They listen and reframe the problem. It is this enseble of the problem framing, on-the-spot experiment, detection of consequences and implcations, back talk and response to back talk, that constitutes a reflective conversation with the materials of the situation – the designlike artistry of professional practice. (pp. 157–158)

From Schön's submission we see that the practitioner, the teacher in our case immerses him/her self into the action of teaching with the tools he/she has got at the moment and then works towards taking control over any emerging real-time occurances. The teacher does this with a conscious mind that is ready to adjust and to take on new courses of action as long as they make sense in terms of achieving the intended learning. Since the teacher cannot step back from the action before the end of the planned lesson, he/she must work to find solutions within his/her current interpretive framework of teaching, that is, the limits of his/her knowledge, skills and attributes at the time. However, the experienced practitioner despite being limited by his/her current interpretive frame, will keep track of the different episodes as they occur during real-time teaching, and these later serve as a basis for future actions and changes in practice. This is where we come to the second question, *about how you the teacher can develop or aquire the knowledge, skills and practice dimensions of the aspects listed in Fig. 3.5.*

Ball and Cohen (1999); Barnett and Hodson (2001); Kayima (2022a), among other educators talk about a teacher's knowledge developed in and with practice (see also Chapter 11). This characteristic knowledge concerns the ability to manage the teaching process. Unfortunately, it is not a very direct or simple task for teacher educators to teach these aspects of teacher knowledge in practical or measurable ways, and like other theoretical perspectives about learning, teacher educators teaching these aspects have a limitation to what they can teach practically or theoretically in lectures. You, the teacher-trainee, must play a bigger role in the development of such knowledge, skills and abilities. You will find a myriad of literature on for example students' attitudes and interest (Bolte et al., 2013; Eilks & Hofstein, 2013), effective feedback practices (Chin, 2006; Mortimer & Scott, 2003), use of questions (Kawalkar & Vijapurkar, 2013; Kayima, 2016; Kayima & Jakobsen, 2020; Kayima & Mkimbili, 2019), teacher graduate attributes and skills (Kayima, 2022b), etc. These and other kinds of advice literature, only provide you with a theoretical knowledge base—the ideas, suggestions and or techniques with which you can deal with different classroom situations, how to motivate or raise students' interest, etc. Having these theoretical understandings, however, does not guarantee a quick and direct transfer or application of this knowledge to real teaching situations. You will gradually and at times unknowingly make use of these theoretical understandings, and your abilities will be strengthened with your growing teaching experience and interaction with your students and other elements in the school environment. Thus, the theoretical knowledge base from teacher education will serve as a rich reference source to refer to when confronted with teaching uncertainties.

In one empirical study, a chemistry teacher trainee was asked after teaching several sessions, to describe what kind of knowledge perspectives from chemistry didactics he had made use of during his several teaching sessions. Below was this response extracted from Kayima (2022a, p. 10).

> On teaching, …. I feel I had more benefit from using social skills and compassion to face situations I did not expect. So, I suppose a combination of regular social skills and the "carrot and stick"-principle (though mostly carrot), works well in complex situations.

While we have no details about what these complex situations were, we assume that these complex situations concern the teacher- students' interactions, involving how students react in different situations based on their own interest versus how the teacher responds back also based on his/her own interest. We can think about this as a situation where the teacher is striving to match or at least try to bring closer his/own interests to those of students in this social learning context. Seen this way, then we come to understand why this teacher trainee highlighted the importance of (social) or personal skills in managing the teaching process.

Kayima (2022a) and Loughran (2002) when talking about effective reflective practice as a key element in developing a teacher's knowledge for teaching, argue for teacher educators to provide *an arena or practical tools* to teacher-trainees that could facilitate teacher-trainees' transitions in thinking about the practice of teaching. One achievable approach suggested is to trigger teacher-trainees' underlying teaching assertions, also often called personal theories about teaching, and let the trainees work towards aligning their initial thinking (personal theories) with research evidenced teaching practices, and experienced real teaching situations. Once trainees are able to forge a relationship between these three aspects (Fig. 3.6), transitions in thinking about practice are made possible and the educator's role will be to negotiate a legitimization of such teaching practices or personal theories that are considered consistent, coherent, results giving, and moreover largely supported by research in science education (Kayima, 2022a).

In Fig. 3.6 we can look at how teachers can gradually change their thinking about the practice of teaching through an ongoing and multiple examination and alignment of their initial thinking with the realities encountered in real practice. Moreover, the pool of knowledge from teacher education serves as a knowledge base against which the legitimacy of experiential knowledge and emerging perspectives is established. Teacher trainees can benefit from having their teacher educators focusing on creating a medium where teacher-trainees make their initial assertions known, discussed and legitimized (Kayima, 2022a).

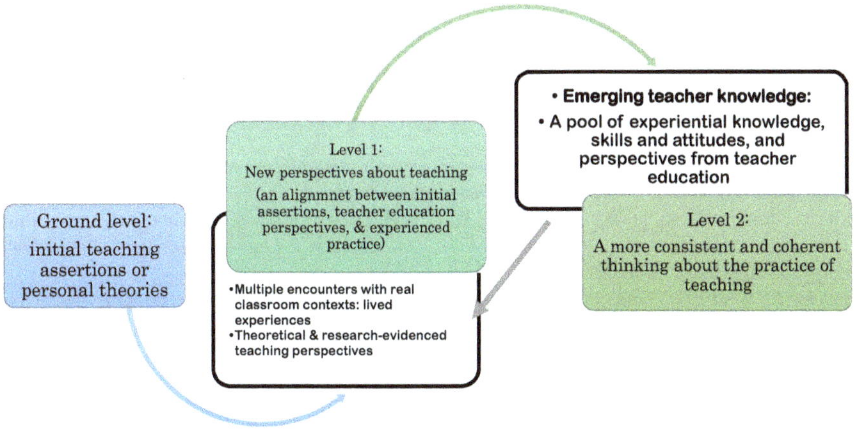

Fig. 3.6 A relation between initial personal theories, experienced practice, and teacher education perspectives

References

Acedo, C., & Hughes, C. (2014). Principles for learning and competences in the 21st-century curriculum. *Prospects, 44*(4), 503–525. https://doi.org/10.1007/s11125-014-9330-1

Akdeniz, C. (2016). Instructional strategies. In C. Akdeniz (Ed.), *Instructional process and concepts in theory and practice: Improving the teaching process* (pp. 57–105). Springer Singapore. https://doi.org/10.1007/978-981-10-2519-8_2

Ball, D. L., & Cohen, D. K. (1999). Developing practice, developing practitioners: Toward a practice-based theory of professional education. In G. Sykes & L. Darling-Hammond (Eds.), *Teaching as the learning profession: Handbook of policy and practice* (Vol. 1, pp. 3–32). Jossey Bass.

Barnett, J., & Hodson, D. (2001). Pedagogical context knowledge: Toward a fuller understanding of what good science teachers know. *Science Education, 85*(4), 426–453. https://doi.org/10.1002/sce.1017

Bobbitt, F. J. (1918). *The curriculum.* University Press.

Bolte, C., Streller, S., & Hofstein, A. (2013). How to motivate students and raise their interest in chemistry education. In I. Eilks & A. Hofstein (Eds.), *Teaching Chemistry—A studybook: A practical guide and textbook for student teachers, teacher trainees and teachers* (pp. 67–95). SensePublishers. https://doi.org/10.1007/978-94-6209-140-5_3

Bucat, R. (2004). Pedagogical content knowledge as a way forward: Applied research in chemistry education [10.1039/B4RP90025A]. *Chemistry Education Research and Practice, 5*(3), 215–228. https://doi.org/10.1039/B4RP90025A

Chin, C. (2006). Classroom interaction in science: Teacher questioning and feedback to students' responses. *International Journal of Science Education, 28*(11), 1315–1346. https://doi.org/10.1080/09500690600621100

Chin, C. (2007). Teacher questioning in science classrooms: Approaches that stimulate productive thinking. *Journal of Research in Science Teaching, 44*(6), 815–843. https://doi.org/10.1002/tea.20171

Chinn, C. A., & Iordanou, K. (2023). Theories of learning. In N. G. Lederman, D. L. Zeidler, & J. S. Lederma (Eds.), *Handbook of research on science education* (pp. 89–120). Routledge Taylor & Francis Group.

Duschl, R. A., & Osborne, J. (2002). Supporting and promoting argumentation discourse in science education. *Studies in Science Education, 38*(1), 39–72. https://doi.org/10.1080/03057260208560187

Eilks, I., & Hofstein, A. (2013). *Teaching chemistry–a studybook: A practical guide and textbook for student teachers, teacher trainees and teachers*. Springer.

Eilks, I., Rauch, F., Ralle, B., & Hofstein, A. (2013). How to allocate the chemistry curriculum between science and society. In I. Eilks & A. Hofstein (Eds.), *Teaching chemistry–A studybook* (pp. 1–36). Sense Publishers.

Geddis, A. N. (1993). Transforming subject-matter knowledge: The role of pedagogical content knowledge in learning to reflect on teaching. *International Journal of Science Education, 15*(6), 673–683. https://doi.org/10.1080/0950069930150605

Gibney, D. (2018). Towards an ideal chemistry curriculum. *School Science Review, 100*(370), 30–35.

Glatthorn, A. A., Boschee, F., Whitehead, B. M., & Boschee, B. F. (2018). The nature of curriculum. In A. A. Glatthorn, F. Boschee, B. M. Whitehead, & B. F. Boschee (Eds.), *Curriculum leadership: Strategies for development and implementation*. SAGE publications.

Goodlad, J. I., Klein, F. M., & Tye, K. A. (1979). The domains of curriculum and their study. In J. I. Goodlad (Ed.), *Curriculum inquiry: The study of curriculum practice* (pp. 43–76). McGraw Hill.

Harris, A. (1998). Effective Teaching: A review of the literature. *School Leadership & Management, 18*(2), 169–183. https://doi.org/10.1080/13632439869628

Jay, P. (2005). *The art of teaching* [Book]. Oxford University Press. https://search.ebscohost.com/login.aspx?direct=true&db=nlebk&AN=140871&site=ehost-live&scope=site

Kawalkar, A., & Vijapurkar, J. (2013). Scaffolding Science Talk: The role of teachers' questions in the inquiry classroom. *International Journal of Science Education, 35*(12), 2004–2027. https://doi.org/10.1080/09500693.2011.604684

Kayima, F. (2016). Question classification taxonomies as guides to formulating questions for use in chemistry classrooms. *European Journal of Science and Mathematics Education, 4*(3), 353–364. https://doi.org/10.30935/scimath/9476

Kayima, F. (2022a). The role of reflective practice in mediating development of pre-service science teachers' professional and classroom knowledge. *Interdisciplinary Journal of Environmental and Science Education, 18*(1), e2262. https://doi.org/10.21601/ijese/11364

Kayima, F. (2022b). Teaching graduate Attributes along with subject content: Perspectives from science teacher educators. *Nordic Studies in Science Education, 18*(3), 290–304. https://doi.org/10.5617/nordina.8982

Kayima, F., & Jakobsen, A. (2020). Exploring the situational adequacy of teacher questions in science classrooms. *Research in Science Education, 50*(2), 437–467. https://doi.org/10.1007/s11165-018-9696-9

Kayima, F., & Mkimbili, S. T. (2019). How do chemistry teachers deal with students' Incorrect/Undesired responses to oral classroom questions? Exploring effective feedback practices [journal article]. *Research in Science Education*. https://doi.org/10.1007/s11165-019-09876-y

Kayima, F., & Mkimbili, S. T. (2021). How do chemistry teachers deal with students' Incorrect/Undesired responses to oral classroom questions? exploring effective feedback practices. *Research in Science Education, 51*(2), 647–668. https://doi.org/10.1007/s11165-019-09876-y

Klein, M. F. (1991). *The politics of curriculum decision-making: Issues in centralizing the curriculum*. Suny Press.

Kousa, P., Kavonius, R., & Aksela, M. (2018). Low-achieving students' attitudes towards learning chemistry and chemistry teaching methods [10.1039/C7RP00226B]. *Chemistry Education Research and Practice, 19*(2), 431–441. https://doi.org/10.1039/C7RP00226B

Krapp, A., & Prenzel, M. (2011). Research on interest in science: Theories, methods, and findings. *International Journal of Science Education, 33*(1), 27–50. https://doi.org/10.1080/09500693.2010.518645

Lang, H. R., & Evans, D. N. (2006). *Models, strategies, and methods for effective teaching*. Pearson/Allyn and Bacon. https://books.google.no/books?id=7XruAAAAMAAJ

Leach, J., & Scott, P. (2003). Individual and sociocultural views of learning in science education. *Science & Education, 12*(1), 91–113. https://doi.org/10.1023/A:1022665519862

Lederman, N. G., & Abell, S. K. (2007). *Handbook of research on science education*. Lawrence Erlbaum Associates.

Lederman, N. G., & Abell, S. K. (2014). *Handbook of research on science education, volume II* (Vol. 2). Routledge.

Loughran, J. J. (2002). Effective reflective practice: In search of meaning in learning about teaching. *Journal of Teacher Education, 53*(1), 33–43. https://doi.org/10.1177/0022487102053001004

Ma, N., Jia-Hui, G., & Sun, Y.-F. (2025). Evaluating scientific argumentation ability: A framework based on scientific argumentation maps. *International Journal of Science Education*, 1–20. https://doi.org/10.1080/09500693.2025.2462796

Marzano, R. J., & Brown, J. L. (2009). *A handbook for the art and science of teaching*. ASCD.

McBrien, J. L., & Brandt, R. S. (1997). *The language of learning: A guide to education terms*. Association for Supervision and Curriculum Development.

Morcke, A. M., Dornan, T., & Eika, B. (2013). Outcome (competency) based education: An exploration of its origins, theoretical basis, and empirical evidence. *Advances in Health Sciences Education, 18*(4), 851–863. https://doi.org/10.1007/s10459-012-9405-9

Mortimer, E., & Scott, P. (2003). *Meaning making in secondary science classroomsaa*. McGraw-Hill Education (UK).

Mullis, I. V. S., Martin, M. O., Goh, S., & Cotter, K. (2016). *TIMSS 2015 Encyclopedia: Education Policy and curriculum in mathematics and science*. TIMSS & PIRLS International Study Center. http://timssandpirls.bc.edu/timss2015/encyclopedia/

Nasiba, H., Bunyod, A., & Ravshan, K. (2021). The main differences between teaching approaches, methods, procedures, techniques, styles and strategies. *JournalNX, 7*(02), 371–375.

Noémie, L. D., Pablo, F., & Guillaume, B. (2016). *Teaching strategies for instructional quality*. https://www.oecd-ilibrary.org/content/paper/5jln1hlsr0lr-en

Orlich, D. C., Harder, R. J., Callahan, R. C., Trevisan, M. S., & Brown, A. H. (2012). *Teaching strategies: A guide to effective instruction*. Cengage Learning.

Osborne, J., Erduran, S., & Simon, S. (2004). Enhancing the quality of argumentation in school science. *Journal of Research in Science Teaching, 41*(10), 994–1020. https://doi.org/10.1002/tea.20035

Özdem Yilmaz, Y., Jale, C., Hamide, E., & Erduran, S. (2017). The pedagogy of argumentation in science education: Science teachers' instructional practices. *International Journal of Science Education, 39*(11), 1443–1464. https://doi.org/10.1080/09500693.2017.1336807

Saskatchewan-Education. (1991). *Instructional approaches: A framework for professional practice*.

Schön, D. A. (1983). *The reflective practioner*. Basic Books.

Schön, D. A. (1987). *Educating the reflective practitioner*. Jossey-Bass.

Soare, E. (2015). Perspectives on designing the competence based curriculum. *Procedia—Social and Behavioral Sciences, 180*, 972–977. https://doi.org/10.1016/j.sbspro.2015.02.259

Sweller, J., & Chandler, P. (1994). *Why some material is difficult to learn* [185–233]. Hillsdale.

Treagust, D. F. (2007). General instructional methods and strategies. In N. G. Lederman & S. K. Abell (Eds.), *Handbook of research on science education* (pp. 373–391).

Treagust, D. F., & Tsui, C.-Y. (2014). General instructional methods and strategies. In N. G. Lederman & S. K. Abell (Eds.), *Handbook of research on science education* (Vol. 2, pp. 303–320). Routledge.

Vural, L. (2016). Instructional methods. In C. Akdeniz (Ed.), *Instructional process and concepts in theory and practice: Improving the teaching process* (pp. 107–145). Springer Singapore. https://doi.org/10.1007/978-981-10-2519-8_3

Yoder, R. J., Bobbitt-Zeher, D., & Sawicki, V. (2021). Understanding the use of student-centered teaching methods in undergraduate chemistry courses. *Research in Science Education, 51*(2), 845–863. https://doi.org/10.1007/s11165-019-9820-5

4 The Role of Students' Misconceptions and Alternative Conceptions in the Learning of Chemistry

Abstract

Construction of new understandings is both influenced and shaped by one's initial (prior)—understandings, past and current experiences, perceptions or beliefs. These constitute a person's interpretive frame and a basis upon which the individual is able to explore and understand the world around him/her. This chapter addresses the role of students' misconceptions and alternative conceptions in the learning of chemistry. Students' initial (prior) understandings are positioned and discussed as key enablers of chemistry learning and how teachers can build upon these to effect conceptual change and foster deep learning of chemistry concepts.

4.1 Students' Initial Understandings—A Foundation for New Learning

Construction of new understandings is both influenced and shaped by one's initial (prior)—understandings, past and current experiences, perceptions or beliefs. These constitute a person's interpretive frame and a basis upon which the individual is able to explore and understand the world around him/her. Regardless of our age, we humans are in a constant search for meaning, for reasons and justifications for all the changes that take place around us, including finding timely answers and solutions to problems and questions arising from our continuous interaction with the surroundings. There is not a point in our lifetime where we do not have an explanation for our observations, actions or for why things around us behave as they do. To put it simply, humans regardless of their age cannot rest until they have some kind of answers, explanations, or a form of justification for whatever is happening around them at anyone moment. Whether or not these answers and justifications are correct or not, verifiably true or not, whether they meet expected societal standards or not, whether they align with scientific understandings or not,

F. Kayima, *Preparing for Chemistry Teaching*, Springer Texts in Education,
https://doi.org/10.1007/978-3-031-97122-8_4

is a question that individuals continuously address as they continue to allow new experiences to shape their understandings. This indisputable truism underscores the central role of a person's interpretive frame (a set of initial understandings, experiences, perceptions and beliefs) in shaping how we understand and interact with our surroundings today, and how today's understandings and experiences help shape how we will explore, understand and interact with our surroundings tomorrow.

In educational settings, each student (or learner), when they come to class (or any other learning arenas/contexts), come with a set of initial understandings, experiences, perceptions or beliefs. This is what they have, and it is where new learning should start from and build or progress further. As part of these initial understandings are those ideas and conceptions that align with the understandings described in science textbooks, and those which are inconsistent with scientific views. The students' initial understandings may be completely deviating from the considered scientific view (principal or concept), or they may be simply incomplete, or wrong explanations for correctly observed phenomena. They might be explanations that seem to make sense to correctly observed phenomena, but which explanations would otherwise be different if the student making these explanations had had access to the correct, acceptable or sufficient scientific principles and explanations. To capture these different elements of a student's initial understandings, different educators have used several terms in a number of educational research reports. Some examples include terms like, students' alternative conceptions (e.g.; Colburn, 2009; Coll & Taylor, 2001; Horton, 2007; Salame et al., 2022; Taber, 2019; Talanquer, 2006), students' misconceptions (e.g.; Barke et al., 2008, 2009; Nahum et al., 2004; Olde Bekkink et al., 2016; Özmen, 2004; Patil et al., 2019; Taber, 2009), Naïve theories or spontaneous conceptions (Mestre & Touger, 1989), misunderstanding (Perkins & Simmons, 1988), erroneous concepts (Bütüner & Filiz, 2017), preconceptions (Benson et al., 1993), prescientific conceptions (Al-Kunifed, 1993), alternative framework (Driver & Easley, 1978; Nussbaum & Novick, 1982), spontaneous knowledge (Horton, 2007), etc.

There are two main sources for students' inherent initial understandings, experiences, perceptions or beliefs, that is, (a) experiences acquired or developed from and within the school environment and (b) those acquired outside the school setting, that is, from our day-to-day interaction with the environment outside the school system. Moreover, these initial understandings stretch beyond specific subject content-related understandings and include accounts of democratically and culturally informed practices, experiences, beliefs and norms. In teaching and learning environments however, interest is more in subject-related initial understandings, experiences, beliefs or perceptions that influence and thereby shape our further understanding of canonical knowledge. It is this latter that is the focus of this chapter.

If we employ a characterization similar to one commonly found in science education literature, then we have, *(i) initial understandings that students come with to the classroom, which have been shaped and developed independent of the classroom teacher's teaching, but through the student's interaction with his/her surroundings over time.* Some scholar describe these as students' preconceptions—"ideas developed without having any prior knowledge of the subject" (Barke et al., 2009, p. 21) or alternative conceptions—"conceptualizations entertained by people that are alternative to what is considered canonical knowledge" (Taber, 2019, p. 450), and *(ii) initial understandings a student has or exhibits, which mostly result from teaching situations and teaching related contexts.* Scholars conceptualize these as school-made misconceptions—developed through "inappropriate teaching methods and materials" (Barke et al., 2009, p. 21); misconceptions—ideas that are inconsistent with current scientific views (Olde Bekkink et al., 2016; Sari et al., 2023), or—consistent errors in understanding of concepts resulting in inherent difficulties in understanding the topic (Jusniar et al., 2020). For example, some misconceptions are claimed to arise from teachers over simplifying concepts, involved abstract mathematical models and improper description of these models (Sari et al., 2023), the gap between chemical concepts presented and level of cognitive development of student, inappropriate organization of concepts in ways that do not consider learning prerequisites, etc. (see Nahadi et al., 2023).

Regardless of whether the classroom teacher has impacted students' underlying initial understandings or not, these initial understandings are characteristic of one or more of these four features,

i. Ideas and conceptions aligning with science understandings, concepts, principles and theories,
ii. Ideas completely deviating from the considered scientific view (principles, concepts or explanations),
iii. Simply incomplete, or wrong explanations for correctly observed phenomena.
iv. Explanations that seem to make sense to correctly observed phenomena, but which explanations would otherwise be different if the student making these explanations had had access to the correct or sufficient scientific principles and explanations.

4.2 Students' Preconceptions—Ideas Developed Without Having Any Prior Knowledge of the Subject

Nearly 30 years ago (as of 2025), my grandmother and I used to sort eggs collected from a small chicken farm she owned. The sorting was meant to identify and separate out the good, fresh and edible eggs from those that had gone bad and hence not edible. The sorting mechanism we used assumed that (see Fig. 4.1),

> When you hold three eggs A, B and C, such that, while egg A holds firmly on the surface, and your right or left hand holds egg C, let egg B be held freely by the tips of the two other eggs A and C (see Figure 4.1). Once this happens, egg B will start rotating between the tips of egg A and egg C. Rotation was indication that egg B is fresh and thus edible or sellable, while failure to rotate was indication that the egg had gone bad and hence not edible.

Fig. 4.1 Egg B held by the tips of egg A and C

Interestingly, this sorting mechanism worked well for us (me and my grandmother at the time). We managed to sort the eggs accordingly. At times and out of curiosity, I used to crack those eggs that had failed to rotate to establish whether these eggs had indeed gone bad and to my surprise, one way or another, these eggs that had failed to rotate, had something in them that impeded the rotation. For example, some eggs had chicks in them, while for others, the Yolk had broken down and mixed with the albumen. I do not remember at any one point questioning my grandmother's sorting mechanism, and neither did I seek to find another and more better reason or explanation for why some eggs rotated and others did not. However, some years later and after I started and had progressed in secondary school, I started to reflect on and to raise questions about my grandmother's sorting mechanism and to search for alternative explanations behind the egg's rotation. For example, I started thinking that there was the possibility that my grandmother was throwing away so many good eggs simply because they did not rotate. I felt the urge to find an explanation that goes further than to simply accept that eggs that did not rotate had gone bad. Unfortunately, I could not find answers in the classroom since there was no direct topic addressing eggs in the way my grandmother and I did.

The above is a typical example of several conceptual ideas, personal theories and frameworks held by individuals in society which are well aligned with the day-to-day practices of individuals. A similar example, which is commonly known is the addition of a small amount of salt to water to make the water boil faster. We all know that adding a given amount of salt to water will (slightly) raise the boiling point of water and obviously the water takes a bit longer to boil. However, many of us have lived and grown up with the understanding that water boils faster with a slight addition of salt. Moreover, because salted water boils differently, heating more vigorously for example, we were convinced that it indeed boiled faster while in the real sense it did not. Also, since the amount of salt added is small, it was difficult for us to recognize the difference in the rate at which the water boiled, given that the small amount of salt added did not seem to raise the boiling point to a level that could be noticed. Children or learners can hold that view to be true until proven otherwise. Since the understanding suits well within the context in which the idea is operationalized and where there are no alternative experiences to counter the current understanding, the learner cannot be faulted for holding this similar understanding until a new perspective is presented to him/her.

Students' preconceptions—ideas developed without having any prior knowledge of the subject, are a powerful resource that provides a basis for students' learning. Moreover, contrary to what is often reported in literature, for these preconceptions like the two examples provided, learners can quickly adjust their thinking the moment they are presented evidence contrary to the understanding they hold. Since these preconceptions are forged in experience and reinforced by associated actions, they can be discarded by learners once evidence is presented that counteract the held views. The teacher's important job therefore is to support students by presenting teaching situations and strategies that could make students reveal their initially held views or at least recognize and reflect on their views even if they do not make them known to the whole class.

Discovering what ideas students hold about a certain scientific concept is not going to be easy. Perhaps you (the teacher) can start by sharing some of your experiences, pointing to some common views that students can associate or identify with. This is likely to encourage them to recognize, reflect and share their own views about the concept being discussed. It is also possible for the teacher to design and use diagnostic questions. However, these will be more successful if they are narrowed down to a concrete topic, and with carefully formulated distractors that help reveal how students view and understand certain elements of the explored topic or concept.

There is a good pool of educational resources with suggestions about how teachers could discover and deal with students' preconceptions which are inconsistent with the scientific concepts of interest. Some examples of these suggestions include (i) teachers' noticing, the ability to identify what is important in a teaching situation, to use what one knows about the context to reason about a situation and to make connections between specific events and broader principles of teaching and learning (Schwichow et al., 2022; van Es & Sherin, 2008), (ii) using analogies and models in instruction, where the analogy is introduced, anchored to a particular contextualized problem to arouse students' held views and intuitions, and then bridging the analogy and the anchor (Clement & Brown, 2008), (iii) teachers' ability to anticipate (Chiu & Lin, 2005; Tanner & Allen, 2005), (iv) using visual representations (Canlas, 2021), (iv) use of a two-tier, multiple choice diagnostic test to identify misconceptions (Nahadi et al., 2023; Tan et al., 2005), use of a standard chemistry misconception test (Mondal & Chakraborty, 2013), etc.

Presenting teaching situations that arouse students' discussions and engagement in argumentation, teachers sharing their own experiences about a phenomenon to arouse students and encourage them to share their own, are some examples of teachers' noticing strategies and ways to anticipate students' preconceptions. Anticipation is possible since teachers themselves have a rich pool of experiences from their childhood that they can use as entry points. Moreover, educational research has documented some of these preconceptions and hence there is a good basis for a teacher to anticipate the preconceptions associated with the topic of interest, and design learning activities that provide a platform for students to share and discuss their preconceptions. Clement and Brown's (2008) use of analogies and models in instruction is about targeting students' preconceptions or intuitions consistent with science concepts and to build students' understanding further by starting with what they know correctly, while simultaneously helping them recognize those intuitions which are inconsistent with science concepts. The use of analogies and models thus is one methodological approach a teacher can employ, having correctly anticipated the preconceptions related to the science concepts of interest.

Before going over to discuss students' school-made misconceptions in the next subsection, what do you think is the correct reasoning behind the rotation of the egg when held between end tips of two other eggs as seen in Fig. 4.1?

4.3 School -Made Misconceptions—Developed Through Inappropriate Teaching Methods, Teaching Materials, or the Nature and Abstractness of Content

4.3.1 Learning from What Chemistry Education Research Has Found and Reported

Let us start with the widespread conception that chemistry is a difficult subject to many students (e.g.; Ayas & Demirbas, 1997; Barrow, 1994; Carter & Brickhouse, 1989; Colburn, 2009; Coll & Taylor, 2001; Gabel, 1998; Nakhleh, 1992; Symington & Kirkwood, 1996). It is widely argued that chemistry as a discipline, comprises many abstract and frequently counter-intuitive concepts (Gabel, 1998), and that it is difficult for students to relate chemical ideas, concepts, principles and theories to their own individual experiences (Colburn, 2009; Taber, 2001). Moreover, in addition to being abstract, chemistry educators and researchers also express that the molecular and symbolic representation dimensions of chemistry are far removed from everyday life. This truism leads to the conclusion shared by some chemistry researchers (e.g.; Nahum et al., 2004; Taber, 2001) that in chemistry, most of the students' initial understandings, ideas or call them misconceptions or alternative conceptions, are developed from and or result from their earlier chemistry or science teachings. The justification for this is that, making sense of abstract notions mainly at the molecular and representation levels of chemistry requires a framework structure (a body of initial knowledge and experiences), which could only be available to students from prior chemistry teachings.

We can thus consider the main source for students' misconceptions, or alternative conceptions in chemistry to be related to.

i. The nature and language of chemistry,
ii. Language, communication and text in chemistry textbooks,
iii. The teacher's knowledge of chemistry, and chemistry teaching competence.
iv. Students' failure to understand that words assume different meanings in different contexts (e.g., similar words from everyday language assume different meanings when used in chemistry).

First, students must have a body of initial knowledge and experiences to relate to if they are to construct new understandings or make sense of new information. As noted earlier, in chemistry, this initial basis knowledge only comes from previous chemistry or science teachings. Knowledge and experiences acquired from earlier chemistry lessons serve as students' initial understandings which they take into future chemistry lessons. Thus, our understanding of students' development of alternative conceptions relies heavily on understanding what happens or has happened in these "earlier" or previous chemistry lessons. This does not imply however that initial understandings (preconceptions) acquired outside teaching contexts do not necessarily play a part in shaping and reinforcing alternative conceptions in chemistry. They indeed do. However, since most of what is learned about and what constitutes chemistry as a subject is far removed from the day-to-day interactions in terms of the commonly used language, symbols and objects, these pre-conceptions would have but less impact.

Regarding the "nature and language of chemistry" as a source for students' alternative conceptions or misconceptions, let us consider that, when confronted with a task of understanding chemical concepts that require students to think and navigate through the macro, molecular and the symbolic levels of chemistry simultaneously or within a short-time lapse, students tend to devise a survival mechanism or a working framework (interpretive framework structure) with which to confront the problem or task at hand. In the process of developing this working framework, they start to search for relations and proximate experiences within their cognitive structures, which could act as anchoring points for the new ideas, knowledge or experiences being digested or processed. While in most cases the complexity and abstractness of the concepts being taught plays a role, this in part depends on how the teacher chooses to unpackage the chemistry concepts to the students. It is also much a question of whether the teacher has a correct understanding of the chemistry concepts, knowledge about the teachability and learnability of those specific concepts or topic s/he intends to teach. For example, when a teacher introduces acids as substances with a sour taste and bases as substances with a bitter taste, such a formulation is homecoming to the student, since there is already an existing pool of proximate experiences to this formulation within the student's cognitive structure. However, if the teacher instead talked about acids and bases in terms of electron donation and acceptance or in terms of oxidation or reduction, the same student would have to respond to this kind of information in a different way, especially if there are no existing pool of proximate experiences within the student's cognitive structure. As an alternative way out, the student then will start to look for common threads within the teacher's formulation to work as anchoring points. In this case, these could be the words "donation" and "acceptance". The student could locate these two words from his existing cognitive structure and hence use them as a starting point for grasping the distinction between acids and bases. Once this is achieved, the student may not continue to strive to understand why and how this understanding comes about.

To further illustrate the above understanding, let us consider and analyze the following misconceptions reported in chemistry education research (Colburn, 2009, p. 10).

i. Atoms are hard spheres, similar to billiard balls.
ii. Atoms are soft, like clouds—like the electron "clouds".
iii. Electrons revolve around nuclei, like planets around the sun.
iv. Molecules literally expand—get bigger—when they are heated,
v. Atoms or molecules are glued together when chemically bonded.
vi. Positive and negative ions together form a molecule when ionically bonded.
vii. Molecules of solids are hard, those of gases are soft, metal atoms are shiny, water molecules are wet.

Reading through the list above, the words (terms) atoms, electrons, ions and molecules are not common in the students' everyday language and interaction. These are words (terms) students first encounter in educational (teaching) settings. When meeting them for the first time, these words will be new additions to the students' cognitive structure or the interpretive frames. The question we must ask ourselves then concerns how the student will meet these words for the first time. Quite often, teachers introduce atoms by expressing that, atoms "are the smallest indestructible particles of matter", and that "matter is anything that occupies space and has weight", and that "all things living and non-living, are matter". With this information at the student's disposal, s/he processes the information, making logical connections between "matter" which represents what is visible (and fathomable) and the "atom" which is the smallest indestructible particle of matter. This student's processing involves the use of the current experiences in the cognitive structure, a process resulting in development of own mental models (Chinn & Iordanou, 2023; Taber, 2001) as the guiding framework for conceptualizing and making sense of the information or task at hand. The process of developing mental models or alternative frameworks is enabled by the student being able to find anchoring points from both his existing cognitive structure and from the familiar language, words or terms presented as part of the task or information being processed. For example, I have observed teachers who bring apples to the classroom and cut them into very small pieces as an introduction to what an atom is. Others use chalk and break these into very small pieces, aiming at a point where they cannot break more pieces from these. What is often forgotten is that the student is continuously processing and working on the information being provided by the teacher in real time, and it becomes very easy to consider the teacher's demonstration as the point of departure for synthesizing and developing a framework for understanding what an atom is?

In view of the above, when looking at the list of misconceptions above, you start to appreciate the substantial effort students put into the process of getting to understand these chemical concepts. Note that there is a distinction between misconceptions developed by the students themselves as they try to understand the chemical concepts presented to them, and misconceptions directly coming from the teacher, who teaches or presents chemical concepts incorrectly (e.g.; Nahadi et al., 2023; Sari et al., 2023). All the listed misconceptions above seem to reflect students' own effort to conceptualize, understand, and communicate these chemical concepts to others in the way they understand and think about them. Considered this way, these understandings reflect a work in progress, and students will be willing to accommodate the correct explanations or understandings as long as their incorrect understandings are discovered, and the right support provided. This however does not mean overlooking the fact that some students are limited in terms of abstract thinking abilities (Herron, 1975), and will have difficulties interpreting and grasping chemical concepts and terms (Erman, 2017; Kozma & Russell, 1997; Krajcik, 1991). Chemistry teachers, however, should not use this argument to justify some students' lack of progress in certain areas of chemistry learning. This thinking itself could serve as a learning block for the students, and teachers should focus on finding ways to support students identified as facing this challenge.

4.4 The Chemistry Teacher: What S/he Teaches and How S/he Teaches

Considering misconceptions directly reinforced by the teacher, by what s/he teaches or through how they teach, the oversimplified way in which chemical concepts are presented to students, results in students overlooking the cognitive investment needed to derive at the correct understandings. Indeed, the teachers' rationale behind presenting chemical concepts in very simplified forms is obviously to help their learners to understand chemical concepts easily. However, in most cases, doing so ends up even making it more difficult for students to understand the concepts correctly.

Taking bonding as one example, quite often, and as an entry into teaching about chemical bonding, teachers find it easier to tell their students that it is very easy for an atom with 7 electrons in its outer most shell to obtain (accept) the one remaining electron to gain a noble gas structure, and for an atom with 1 electron in its outer most shell to donate that electron and retain a stable structure. This seemingly working rule of thumb tends to make it easier for both teacher and students when they talk about the electron configuration, ionic and covalent bonding, using the octet rule as an explanatory principle. However, the way the octet rule is conceptualized and used results in students' wrong conceptual understanding of the underlying principles of bonding. By using the octet-rule as a gateway into the teaching of chemical bonding, teachers tend to overlook the other but central factors responsible for bonding, such as ionization energy and electron affinity or generally the role of electrostatics and energetics in determining bonding. Moreover, some teachers fail to emphasize to their students that the octet-rule is simply a summary of what has been observed and that not all molecules (ions or compounds) obey the octet rule in terms of the way they are organized. As such, students tend to believe that (i) bonding is only about electron sharing between nonmetals and electron transfer from a metal to a nonmetal, and that (ii) the main cause of chemical bonding is for atoms to get their outermost shells full. Moreover, students tend to consider the number of ionic bonds formed to be determined by the number of electrons donated or accepted and not just the ionic charges (see; Lee, 2007; Taber, 2003). It is thus no surprise when research findings for example, reveal that some students have difficulties understanding what kind of bonding exists in metals, or how bonding in lattice structures such as NaCl could be explained with respect to their initial understanding that is guided by the octet rule.

In addition to presenting chemical concepts in very simplified forms, a teacher's lack of profound knowledge about the teachability and learnability of individual chemical concepts (topics) and the variations that exist from topic to topic largely contribute to students' misunderstanding of intended concepts. Let us think about the "teachability and learnability" of a given chemistry topic or concept as having a very close friend whom you know very well and understand inside-out. This pedagogical element of a teacher's knowledge for teaching, entails a knowledge that informs about the important aspects of the content to be taught, how to re-package the content into forms consumable by learners, and how to present and communicate the content to learners (Bucat, 2004, 2005; Geddis, 1993). Note that these competence demands vary from topic to topic, that is, the repackaging and content breakdowns needed for teaching and learning about redox reactions are not necessarily the same when the topic of study is about rates of reaction and so forth. Teachers themselves need to understand chemical concepts correctly if they are to teach these concepts correctly to their students. There are however several research reports that indicate that teachers have misconceptions themselves in several chemistry topics (Erman, 2017; Kolomuç & Tekin, 2011). In this case, then it will be difficult to teach their students correctly when they themselves do not have the correct understanding. More still, for teachers with sufficient and correct understanding of the chemistry concepts, this knowledge alone is not sufficient. There is a need for the teacher to communicate effectively with students, navigating the chemistry language used to talk about concepts and the coherent relation between several concepts, the terms and symbols used. Teachability and learnability also entails pedagogically being aware of the interdependence of chemical concepts and terms used in chemistry, and how the unthought-through and illogical use of the chemistry language, terms and symbols, could impede students' further understanding.

Before I invite you to reflect on why knowledge about students' misconceptions is important to you as a teacher, I want to point to the impression obtained by some teachers, and which you may also get when you read conclusions from some published research articles and books about students' alternative conceptions and misconceptions. When you consider some published reports addressing this aspect of students' misconceptions and preconceptions, it is easy to get the impression that these students' initial understandings largely impede students' learning rather than providing a basis for learning. Indeed, this negative picture seems to dominate discussions about students' alternative conceptions and misconceptions in science teaching. More emphasis and weight are put on what the students get right or wrong, and less on exploiting students' initial understandings as a basis for negotiating and building new understandings. On the contrary, the presentation and examples in this chapter should provide you with the opportunity to rethink and adjust the way you look at and consider what students bring with them into the classroom and plan your teaching in a way that takes advantage of and make use of students' preconceived understandings.

Moreover, you should assess your own chemistry knowledge, considering the possibility that you might yourself have some chemical concepts you might have understood incorrectly. An effective teacher should recognize the limits of his/her understanding when addressing a certain topic to avoid misconceptions that can be created by introducing and explaining those concepts for which s/he lacks sufficient understanding. It is better to postpone, do sufficient preparation and then take up the topic again when fully prepared. Recall that I earlier noted that sufficient preparation or having good knowledge of the content does not necessarily translate into good or effective teaching. A teacher needs to learn from earlier experiences that have been made known through published educational resources. As mentioned, already, most of the common misconceptions in chemistry have been documented (e.g.; Kimbrough & Jensen, 2009; Majeed et al., 2023; Mondal & Chakraborty, 2013). This is a good starting point. Find and read about these documented cases of misconceptions and what suggestions have been presented regarding how to deal with the reported misconceptions and make attempts to plan and develop a teaching plan where you try out some of those suggested recommendations.

Finally, both knowledge about existing students' misconceptions and the possible ways to address them in teaching is important. However, when we think only in terms of addressing misconceptions, we might be forced to think that students holding misconceptions or alternative conceptions is something that is strongly negative. On the contrary, and as has been repeatedly emphasized throughout the chapter, both students' preconceptions and misconceptions are part of a knowledge students draw on to propel their learning. They are the anchoring points for new learning and hence teachers should and are expected to draw on these to foster students' learning.

4.5 Learning and Relearning: Effecting Conceptual Change in Chemistry

To this end, there is no doubt to the fact that learners come to our chemistry classrooms with both pre-instructional knowledge structures and those knowledge structures that have developed, been shaped and reinforced through their previous chemistry (science) learning encounters, that is, misconceptions resulting from the textbooks, instructional materials, educational analogies, teachers' misconceptions, and teaching methods among other sources. We have also come to understand and as documented in educational literature, that any science instruction characterized to be effective, must consider these pre-conceptions and misconceptions (e.g.:, Addido et al., 2022; Anggoro et al., 2019; Bigozzi et al., 2018). Thus, we are expected to support students to acquire new understandings through a process that involves both learning and relearning (Acedo & Hughes, 2014), a process that involves deep conceptual understanding that provides for conceptual change. That is, through a process that allows students to reflect about and express what they already know about the concepts being explored and draw on these prior understandings to mediate and engage in efforts to derive at new or improved understandings together with their teacher. Consequently, questions relating to how teachers can deal with students' misconceptions and scientifically incorrect pre-instructional knowledge structures can be answered by considering understandings about conceptual change presented in science education research.

Efforts to help students obtain correct or acceptable understandings of science concepts have been ongoing since the early 1980s. Since the proposition of a Conceptual Change Model (CCM) by Posner et al. (1982), there has been a growing pool of suggestions in (science) education concerning how conceptual change occurs and about plausible instructional approaches and strategies for effecting conceptual change in science. Pacaci et al. (2024) describe three different conceptual change instructional strategies in their meta-analysis of several experimental studies, which respectively derive from three different perspectives about the nature of misconceptions and plausible pathways to effect conceptual change. First is *the cognitive conflict strategy* that is premised on Posner et al.'s (1982) conceptual change model (CCM). The cognitive conflict instructional strategy involves (i) students becoming aware of their inconsistencies between their prior knowledge and the scientific knowledge in question, and become dissatisfied, (ii) they are presented with new knowledge, ideas or concepts that they should find intelligible, (iii) the presented knowledge, ideas or concepts should be plausible, consistent and should have the capacity to solve problems, and (iv) the presented knowledge, ideas or concepts should be fruitful in terms of presenting new insights and or discoveries.

Presenting dissonance between what students believe and what they can observe is often employed by teachers in chemistry classrooms. Take for example using the *cognitive conflict strategy* to address misconceptions regarding conservation of mass during combustion. Several misconceptions related to students' understanding of conservation of mass during combustion are reported (see for example;

Barke et al., 2009; Barker & Millar, 1999; Basheer et al., 2018; Tokgöz & Sarıkaya, 2025). Examples reported include students having difficulties understanding the conservation of mass and chemical reactions in open and inside a closed system (e.g.; Barker & Millar, 1999; Boujaoude, 1991; Gabel et al., 2001; ÖZmen & Ayas, 2003); distinguishing physical from chemical changes (e.g.:, Eilks et al., 2007; Hesse Iii & Anderson, 1992; Jegstad et al., 2024). For example, students reportedly struggle with differentiating between chemical reaction and physical change, and with the conception of whether gas has mass or not, and with understanding the role of oxygen in combustion reactions (e.g. Boujaoude, 1991; Gabel et al., 2001; Jegstad et al., 2024). A teacher who decides to employ *the cognitive conflict strategy* needs to design appropriate activities that will allow students to expose and make known their initial understandings about the concepts being explored and then present activities that yield evidence that opens for students to start questioning their initial understandings, supported with well-structured, satisfying and justifying explanations or evidence for the new understandings being advanced.

In other words, students experience cognitive dissonance when their predictions fail or when they are exposed to understandings about something in ways contrary to what they believed they knew or how they understood it to be. This presents the teacher with the opportunity to mediate construction of new understandings through discussion and presenting evidence. Among the commonly known examples is the candle combustion lesson where both teacher and students engage in a discussion about where the candle goes as it burns, or what happens to the wax, and then a teacher performs an experiment allowing students to experience cognitive dissonance, which is finally resolved through discussion supported by evidence from the experiment. Other related examples include burning steel wool in both a closed and open system where students get to discuss the role of oxygen in accounting for the mass in the open system, or the role of oxygen in the combustion of wood, etc.

The second conceptual change instructional strategy is termed as or called the ***cognitive bridging strategy*** by Pacaci et al. (2024). It emanates from a conception and characterization of misconceptions that differs from the theory-like and coherent knowledge structure perspective originally advanced by Posner et al. (1982) and later proponents of the cognitive conflict strategy. The ***cognitive bringing strategy*** is premised on the understanding that students' misconceptions are not theory -like and are not coherent but fragmented in nature (diSessa, 1993; diSessa & Sherin, 1998; Hammer, 1996). Scholars subscribing to this perspective consider students' misconceptions to be "large amounts of fragmented knowledge pieces abstracted from every-day experiences" (Pacaci et al., 2024, p. 1266). diSessa (1993) termed these conceived individual's "small cognitive elements" phenomenological primitives (p-prims), ideas "that are not formally learned; that are neither correct nor incorrect themselves, but can be appropriately or inappropriately applied in a given context; and that are activated in concert with other elements." (Harlow & Bianchini, 2020, p. 389). Misconceptions are weakly coherent structures of p-prims, where some of the student's p-prims could be productive and others unproductive depending on the context. Accordingly, new knowledge develops through gradually adding new information or knowledge to existing productive p-prims (diSessa & Levin, 2021; Smith et al., 1993).

With the cognitive bridging instruction strategy, the focus is not about making students dissatisfied with their initial understandings as an entry to effecting conceptual change. Rather, the teacher draws on students' productive p-prims to explain phenomena and to effect new understandings. This understanding of students' misconceptions and their pre-instructional knowledge structures allows us as chemistry teachers to consider these students' initial understandings to be central to their learning of chemistry and science in general. As mentioned earlier, productivity of p-prims is context dependent and hence a teacher can take advantage of these initial understandings to mediate and promote new understandings through combining both cognitivists and constructivists (participationist) teaching strategies. The link between existing and new knowledge is central to achieving conceptual change under the cognitive bridging instruction strategy and so is the use of student-centered instructional approaches. Moreover, the concept of deep conceptual learning (e.g.; Curtis et al., 2002; Hilton & Pellegrino, 2012; Sawyer, 2006) comes to the fore in view of the cognitive bridging instruction strategy. Accordingly, the teacher creates opportunities to allow students to make connections between concepts being explored and between new knowledge and students' prior knowledge. Students are active in their own learning process and are offered opportunities to reflect on their own learning through the metacognitive engagement strategies provided by the teacher. In so doing, students develop well-organized knowledge structures around the main concept of study and can apply the new understandings to new issues within the same knowledge domain.

Moreover, anchoring and bringing analogies (e.g.; Clement, 1993) can be used by teachers to support students in the process of making use of their productive p-prims, digesting new evidence and aligning their initial understandings with the new and correct (acceptable) scientific understandings. Below (Table 4.1) is an example that can be expanded to include more activities that engage students in a dialogic and interactive session to effect conceptual understanding by starting from what they know and building correct concept descriptions with the help of bridging analogies.

The third conceptual change instructional strategy is *the ontological category shift strategy*. This strategy builds upon the understanding of students' misconceptions on "the ontological nature of entities in the world and the nature of scientific conceptions" (Pacaci et al., 2024, p. 1266). Accordingly, conceptual change is conceptualized in terms of ontological shifts (Rusanen & Pöyhönen, 2013), a theoretical conception put forward by Chi and her colleagues (Chi, 1992, 2008; Chi et al., 1994). According to Chi and her colleagues, there are different ontological categories to which entities in the world belong, such as matter, processes, and mental states (Chi & Slotta, 1993), and the process of learning involves organizing objects under the appropriate ontological category (Pacaci et al., 2024; Rusanen & Pöyhönen, 2013). Learning outcomes according to the ontological shifts conceptual change view is considered a "psychological tree of concepts" in the learner's

Table 4.1 Using a bridging analogy between what students know and how they know it and what they should know and how they should know it

Topic	*Bonding*
Subtopic lesson in focus	Ionic bonding
Previously covered concepts	Introduction to the atomic structure, the Bohr's and Schrödinger's atom models, electronic configuration, elements, and compounds, the periodic law and an introduction the periodic system
Topic of discussion	Understanding ionic bonding and exploring the misunderstandings resulting from the simplified conception of the octet rule
Discussion question	Why do members of group 1 elements have to give up their valence electron when they bond with any given halogen in group 17? For example, in the case of sodium and chlorine
Possible misconception	Atoms form bonds because they want to fill their outer electron shells (Luxford & Bretz, 2014; Tsaparlis et al., 2018)
Engage students, discuss their expressed understanding and present justifications that both acknowledge and reframe and provide a new direction of understanding	For example, the idea of "wanting a full shell" as a short cut can be a starting point for you to invite your students to think in more realistic terms, for example in terms of changes in energy, electronegativity and electrostatic attractions. They also understand that atoms do not even possess feelings and desires and wanting a full shell is literal thinking to enable an easy grasp of the idea
Introduce, discuss and engage in a dialogic exchange	What happens between sodium and chlorine as a factor of electronegativity, the need to lower electric potential energy and the factor of electrostatic attractions

(continued)

Table 4.1 (continued)

Topic	*Bonding*
Bridge using analogy:	the "ball rolling downhill" analogy as an example: "Imagine you place a ball on a hill — it rolls down, not because it wants to, but because it's more stable at the bottom. Similarly, atoms don't 'want' anything, but they tend to move toward arrangements with lower energy" "When sodium loses an electron and chlorine gains one, they both end up in lower-energy, more stable states Please note this very explanation could even lead to more misconceptions. Students need to know that the real low energy is achieved when the ionic compound e.g. NaCl, is formed. However, the initial stages of, for example, Ionization of Na costs energy and it is only when the Na ion pairs with chlorine, that the overall system releases more energy compared to the energy invested
Using the descriptions from the analogy in the formulation and working with chemistry concepts	For example, describe the formation of NaCl in terms of energy: Ionization energy of Na Electron affinity of Cl Lattice energy released when oppositely charged ions attract Let students transfer these understandings to establish whether they make sense when bonding between other metal atoms and non-metal atoms are considered

mind, which should be *isomorphic* to the "physical tree" of categories in nature if learning is to be counted as successful (Rusanen & Pöyhönen, 2013, p. 1396). In other words, students need to be made aware of the existing and different ontological categories as well as the ontological category to which certain scientific concepts of interest belong. The need for ontological shift arises when there is an ontological difference between students' conceptualization and the scientific concept of interest. At this point, the student exhibits an ontological category for the target concept though the categorization of concepts is incompatible with the correct ontological category. The teacher then focuses on working with students to *identify* or *assign* a concept to a conceptual category to which it belongs (Chi, 1992, 2008; Chi et al., 1994). If however, the student is identified to lack an appropriate category, then the teacher has to focus attention on helping the student to build and acquire a new category (Pacaci et al., 2024; Rusanen & Pöyhönen, 2013).

For us to be able to understand how we can apply the ontological category shift strategy, we need to understand and be able to distinguish the different ontological categories in chemistry. As mentioned earlier, Chi and her colleagues

(Chi & Slotta, 1993; Chi et al., 1994) considered three ontological category examples, (i) matter (objects, substances or stuff), (ii) processes, and (iii) mental constructs—non-physical models and tools explaining behavior, to which we can add (iv) properties or quantities and (v) structures—arrangements or spatial configurations of matter. Each one of these ontological categories comprises subcategories and what was termed by Chi and her colleagues as ontological attributes. For example, subcategories under matter include pure substances (elements and compounds), mixtures—homogenous and heterogenous mixtures, the macro and microscopic matter. The attributes under matter as an ontological category include, for example, matter occupying space and having weight (mass), being made of particles, existing in different states (liquid, solid, gas and plasma), etc. Several examples of ontological confusion in chemistry are reported including students confusing substances with particles, the idea that gases are not matter since they are invisible, describing chemical reactions as static things, treating energy as a substance, confusing the distinction between atoms, elements and molecules, equating models to reality, misunderstanding scale and composition, e.tc. (for concrete examples see; Chi, 2005; Gilbert & Treagust, 2009; Johnstone, 1991, 2000; National-Research-Council, 2012; Taber, 2002, 2013).

Applying the ontological category shift strategy in your teaching does not demand a completely unique instructional methodology other than employing the commonly known student-centered techniques and deep conceptual learning strategies that engage students. The use of metacognition and multiple representations is pointed out as one way to support ontological category shifts (Gupta et al., 2010); and employing instruction interventions along with category shifting prompts that help students to reframe concepts (Krajcik & Shin, 2023; Slotta et al., 1995). A teacher will need to find ways to elicit and make students' ontological confusion or misconceptions known. Remember, as mentioned, the main goal is not to identify and correct misconceptions but rather to invoke students' pre-knowledge, conceptions, ideas or understandings so that new learning can build upon this existing knowledge structure. As such, you the teacher need to be in a good position to identify these deviating conceptions presented by students and address them systematically. When students express ontological confusion about the concept being studied, the teacher's task is to present the correct or acceptable version of the explored ontological category by making it explicit to the students. Moreover, the nature of the chemical language plays a central role. Students can easily get confused when, for example, analogies, models or other representations are used to abstract complex chemical concepts. For example, using substance-like metaphors—"energy flows into the bond", the bond releases energy, etc., create a totally different conception about the relationship between energy and bonding.

References

Acedo, C., & Hughes, C. (2014). Principles for learning and competences in the 21st-century curriculum. *Prospects, 44*(4), 503–525. https://doi.org/10.1007/s11125-014-9330-1

Addido, J., Burrows, A. C., & Slater, T. F. (2022). Addressing pre-service teachers' misconceptions and promoting conceptual understanding through the conceptual change model. *Problems of Education in the 21st Century, 80*(4), 499–515.

Al-Kunifed, A. A. (1993). *Investigation of high school chemistry students' concepts of chemical symbol, formula, and equation: Students' prescientific conceptions.* Louisiana State University and Agricultural & Mechanical College.

Anggoro, S., Widodo, A., Suhandi, A., & Treagust, D. F. (2019). Using a discrepant event to facilitate preservice elementary teachers' conceptual change about force and motion. *Eurasia journal of mathematics, science and technology education, 15*(8).

Ayas, A., & Demirbas, A. (1997). Turkish secondary students' conceptions of the introductory concepts. *Journal of Chemical Education, 74*(5), 518. https://doi.org/10.1021/ed074p518

Barke, H.-D., Hazari, A., & Yitbarek, S. (2008). *Misconceptions in chemistry: Addressing perceptions in chemical education.* Springer Science & Business Media.

Barke, H.-D., Hazari, A., & Yitbarek, S. (2009). Students' misconceptions and how to overcome them. In *Misconceptions in chemistry: Addressing perceptions in chemical education* (pp. 21–36). Springer Berlin Heidelberg. https://doi.org/10.1007/978-3-540-70989-3_3

Barker, V., & Millar, R. (1999). Students' reasoning about chemical reactions: What changes occur during a context-based post-16 chemistry course? *International Journal of Science Education, 21*(6), 645–665. https://doi.org/10.1080/095006999290499

Barrow, G. M. (1994). General chemistry and the basis for change. *Journal of Chemical Education, 71*(10), 874. https://doi.org/10.1021/ed071p874

Basheer, A., Kortam, N., Zahran, N., Hofestein, A., & Hugerat, M. (2018). Misconceptions among middle school students regarding the conservation of mass during combustion. *Eurasia Journal of Mathematics, Science and Technology Education, 14*(7), 3109–3122.

Benson, D. L., Wittrock, M. C., & Baur, M. E. (1993). Students' preconceptions of the nature of gases. *Journal of Research in Science Teaching, 30*(6), 587–597. https://doi.org/10.1002/tea.3660300607

Bigozzi, L., Tarchi, C., Fiorentini, C., Falsini, P., & Stefanelli, F. (2018). The influence of teaching approach on students' conceptual learning in physics [Original Research]. *Frontiers in Psychology, 9.* https://www.frontiersin.org/journals/psychology/articles/10.3389/fpsyg.2018.02474

Boujaoude, S. B. (1991). A study of the nature of students' understandings about the concept of burning. *Journal of Research in Science Teaching, 28*(8), 689–704. https://doi.org/10.1002/tea.3660280806

Bucat, R. (2004). Pedagogical content knowledge as a way forward: Applied research in chemistry education [10.1039/B4RP90025A]. *Chemistry Education Research and Practice, 5*(3), 215–228. https://doi.org/10.1039/B4RP90025A

Bucat, R. (2005). Implications of chemistry education research for teaching practice: Pedagogical content knowledge as a way forward. *Chemistry Education International, 6*(1), 1–2.

Bütüner, S. Ö., & Filiz, M. (2017). Exploring high-achieving sixth grade students' erroneous answers and misconceptions on the angle concept. *International Journal of Mathematical Education in Science and Technology, 48*(4), 533–554. https://doi.org/10.1080/0020739X.2016.1256444

Canlas, I. P. (2021). Using visual representations in identifying students' preconceptions in friction. *Research in Science & Technological Education, 39*(2), 156–184. https://doi.org/10.1080/02635143.2019.1660630

Carter, C. S., & Brickhouse, N. W. (1989). What makes chemistry difficult? Alternate perceptions. *Journal of Chemical Education, 66*(3), 223. https://doi.org/10.1021/ed066p223

Chi, M. (1992). Conceptual change within and across ontological categories: Examples from learning and discovery in science. In R. Giere & H. Feigl (Eds.), *Cognitive Models of Science* (Vol. 15, pp. 129–186). University of Minnesota Press.

Chi, M. (2008). Three types of conceptual change: belief revision, mental model and categorical shift. *International Handbook of Research on Conceptual Change* (pp. 61–82).

Chi, M. T. H. (2005). Commonsense conceptions of emergent processes: Why some misconceptions are robust. *Journal of the Learning Sciences, 14*(2), 161–199. https://doi.org/10.1207/s15327809jls1402_1

Chi, M. T. H., & Slotta, J. D. (1993). The ontological coherence of intuitive physics. *Cognition and Instruction, 10*(2–3), 249–260. https://doi.org/10.1080/07370008.1985.9649011

Chi, M. T. H., Slotta, J. D., & De Leeuw, N. (1994). From things to processes: A theory of conceptual change for learning science concepts. *Learning and Instruction, 4*(1), 27–43. https://doi.org/10.1016/0959-4752(94)90017-5

Chinn, C. A., & Iordanou, K. (2023). Theories of learning. In N. G. Lederman, D. L. Zeidler, & J. S. Lederma (Eds.), *Handbook of research on science education* (pp. 89–120). Routledge Taylor & Francis Group.

Chiu, M.-H., & Lin, J.-W. (2005). Promoting fourth graders' conceptual change of their understanding of electric current via multiple analogies. *Journal of Research in Science Teaching, 42*(4), 429–464. https://doi.org/10.1002/tea.20062

Clement, J. (1993). Using bridging analogies and anchoring intuitions to deal with students' preconceptions in physics. *Journal of Research in Science Teaching, 30*(10), 1241–1257. https://doi.org/10.1002/tea.3660301007

Clement, J. J., & Brown, D. E. (2008). Using analogies and models in instruction to deal with students' Preconceptions*. In J. J. Clement (Ed.), *Creative model construction in scientists and students* (pp. 139–155). Springer Netherlands. https://doi.org/10.1007/978-1-4020-6712-9_10

Colburn, A. (2009). Alternative conceptions in chemistry. *The Science Teacher, 76*(6), 10.

Coll, R. K., & Taylor, N. (2001). Alternative conceptions of chemical bonding held by upper secondary and tertiary students. *Research in Science & Technological Education, 19*(2), 171–191. https://doi.org/10.1080/02635140120057713

Curtis, P. C., Labov, J. B., Bertenthal, M. W., & Gollub, J. P. (2002). *Learning and understanding: Improving advanced study of mathematics and science in US high schools.* National Academies Press.

diSessa, A. A. (1993). Toward an epistemology of physics. *Cognition and Instruction, 10*(2–3), 105–225. https://doi.org/10.1080/07370008.1985.9649008

diSessa, A. A., & Sherin, B. L. (1998). What changes in conceptual change? *International Journal of Science Education, 20*(10), 1155–1191. https://doi.org/10.1080/0950069980201002

diSessa, A. A., & Levin, M. (2021). Processes of building theories of learning: Three contrasting cases. In O. Levrini, G. Tasquier, T. G. Amin, L. Branchetti, & M. Levin (Eds.), *Engaging with contemporary challenges through science education research: Selected papers from the ESERA 2019 Conference* (pp. 225–236). Springer International Publishing. https://doi.org/10.1007/978-3-030-74490-8_18

Driver, R., & Easley, J. (1978). *Pupils and paradigms: A review of literature related to concept development in adolescent science students.*

Eilks, I., Moellering, J., & Valanides, N. (2007). Seventh-grade students' understanding of chemical reactions: Reflections from an action research interview study. *Eurasia Journal of Mathematics, Science and Technology Education, 3*(4), 271–286.

Erman, E. (2017). Factors contributing to students' misconceptions in learning covalent bonds. *Journal of Research in Science Teaching, 54*(4), 520–537. https://doi.org/10.1002/tea.21375

Gabel, D. (1998). The complexity of chemistry and implications for teaching. In B. J. Fraser & K. G. Tobin (Eds.), *International handbook of science education* (Vol. 1, pp. 233–248). Kluwer.

Gabel, D. L., Stockton, J. D., Monaghan, D. L., & MaKinster, J. G. (2001). Changing children's conceptions of burning. *School Science and Mathematics, 101*(8), 439–451. https://doi.org/10.1111/j.1949-8594.2001.tb17879.x

Geddis, A. N. (1993). Transforming content knowledge: Learning to teach about isotopes. *Science Education, 77*(6), 575–591.

Gilbert, J. K., & Treagust, D. F. (2009). *Multiple representations in chemical education* (Vol. 4). Springer.

Gupta, A., David, H., & Redish, E. F. (2010). The case for dynamic models of learners' ontologies in physics. *Journal of the Learning Sciences, 19*(3), 285–321. https://doi.org/10.1080/10508406.2010.491751

Hammer, D. (1996). Misconceptions or P-Prims: How may alternative perspectives of cognitive structure influence instructional perceptions and intentions. *Journal of the Learning Sciences, 5*(2), 97–127. https://doi.org/10.1207/s15327809jls0502_1

Harlow, D. B., & Bianchini, J. A. (2020). Knowledge-in-Pieces—Andrea A. diSessa, David Hammer. In B. Akpan & T. J. Kennedy (Eds.), *Science education in theory and practice: An introductory guide to learning theory* (pp. 389–401). Springer International Publishing. https://doi.org/10.1007/978-3-030-43620-9_26

Herron, D. J. (1975). Piaget for chemists. Explaining what "good" students cannot understand. *Journal of Chemical Education, 52*(3), 146. https://doi.org/10.1021/ed052p146

Hesse Iii, J. J., & Anderson, C. W. (1992). Students' conceptions of chemical change. *Journal of Research in Science Teaching, 29*(3), 277–299. https://doi.org/10.1002/tea.3660290307

Hilton, M. L., & Pellegrino, J. W. (2012). *Education for life and work: Developing transferable knowledge and skills in the 21st century.* National Academies Press.

Horton, C. (2007). Student alternative conceptions in chemistry. *California Journal of Science Education, 7*(2), 18–38.

Jegstad, K. M., Mestad, I., & Ahmed, S. (2024). Supporting 10th grade students' explanations about mass conservation in combustion reactions. *Revisioning STEAM education in times of climate change*, 351.

Johnstone, A. H. (1991). Why is science difficult to learn? Things are seldom what they seem. *Journal of computer assisted learning, 7*(2), 75–83. https://doi.org/10.1111/j.1365-2729.1991.tb00230.x

Johnstone, A. H. (2000). Teaching of chemistry—logical or psychological? [10.1039/A9RP90001B]. *Chemistry Education Research and Practice, 1*(1), 9–15. https://doi.org/10.1039/A9RP90001B

Jusniar, J., Effendy, E., Budiasih, E., & Sutrisno, S. (2020). Misconceptions in rate of reaction and their impact on misconceptions in chemical equilibrium. *European Journal of Educational Research, 9*(4), 1405–1423.

Kimbrough, D. R., & Jensen, A. C. (2009). Study of common misconceptions in chemistry among middle school science teachers and chemistry professionals. *Abstracts of Papers of the American Chemical Society, 237*. <Go to ISI>://WOS:000207857801421

Kolomuç, A., & Tekin, S. (2011). Chemistry teachers' misconceptions concerning concept of chemical reaction rate. *International Journal of Physics and Chemistry Education, 3*(2), 84–101. https://doi.org/10.51724/ijpce.v3i2.194

Kozma, R. B., & Russell, J. (1997). Multimedia and understanding: Expert and novice responses to different representations of chemical phenomena. *Journal of Research in Science Teaching, 34*(9), 949–968. https://doi.org/10.1002/(SICI)1098-2736(199711)34:9<949::AID-TEA7>3.0.CO;2-U

Krajcik, J., & Shin, N. (2023). Student conceptions, conceptual change, and learning progressions. In G. L. Norman, L. Z. Dana, & S. L. Judith (Eds.), *Handbook of research on science education* (1 ed.). Routledge.

Krajcik, J. S. (1991). Developing student's understanding of chemical concepts. In S. M. Glynn, B. K. Britton, & R. H. Yeany (Eds.), *The psychology of learning science* (Vol. 1, pp. 117–148). Routledge.

Lee, M. (2007). *High school students' alternative ideas about chemical bonding*. D. Foley.

Luxford, C. J., & Bretz, S. L. (2014). Development of the bonding representations inventory to identify student misconceptions about covalent and ionic bonding representations. *Journal of Chemical Education, 91*(3), 312–320. https://doi.org/10.1021/ed400700q

Majeed, S., Ahmad, R., & Mazhar, S. (2023). An exploration of students' common misconceptions in the subject of chemistry at secondary level. *Annals of Human and Social Sciences, 4*(2), 265–272.

Mestre, J., & Touger, J. (1989). Cognitive research-what's in if for physics teachers? *The Physics Teacher, 27*(6), 447–456. https://doi.org/10.1119/1.2342824

Mondal, B. C., & Chakraborty, A. (2013). *Misconceptions in chemistry: Its identification and remedial measures*. Lap Lambert Academic Publishing.

Nahadi, N., Siswaningsih, W., Firman, H., Dewi, E. P., Lestari, T., & Rahmawati, T. (2023). Development and application of a two-tier acid-base misconception diagnostic test based on pictorial to identifying student misconceptions in chemistry. *Journal of Engineering Science and Technology, 18*, 207–223. <Go to ISI>://WOS:001024772800018

Nahum, T. L., Hofstein, A., Mamlok-Naaman, R., & Bar-Dov, Z. (2004). Can final examinations amplify students' misconceptions in chemistry? [10.1039/B4RP90029D]. *Chemistry Education Research and Practice, 5*(3), 301–325. https://doi.org/10.1039/B4RP90029D

Nakhleh, M. B. (1992). Why some students don't learn chemistry: Chemical misconceptions. *Journal of Chemical Education, 69*(3), 191. https://doi.org/10.1021/ed069p191

National-Research-Council. (2012). *A Framework for K-12 science education: Practices, crosscutting concepts, and core ideas*. National Academies Press. https://books.google.no/books?id=b2L5VShktGIC

Nussbaum, J., & Novick, S. (1982). Alternative frameworks, conceptual conflict and accommodation: Toward a principled teaching strategy. *Instructional Science, 11*(3), 183–200. https://doi.org/10.1007/BF00414279

Olde Bekkink, M., Donders, A. R. T. R., Kooloos, J. G., de Waal, R. M. W., & Ruiter, D. J. (2016). Uncovering students' misconceptions by assessment of their written questions. *BMC Medical Education, 16*(1), 221. https://doi.org/10.1186/s12909-016-0739-5

Pacaci, C., Ustun, U., & Ozdemir, O. F. (2024). Effectiveness of conceptual change strategies in science education: A meta-analysis. *Journal of Research in Science Teaching, 61*(6), 1263–1325. https://doi.org/10.1002/tea.21887

Patil, S. J., Chavan, R. L., & Khandagale, V. (2019). Identification of misconceptions in science: Tools, techniques & skills for teachers. *Aarhat Multidisciplinary International Education Research Journal (AMIERJ), 8*(2), 466–472.

Perkins, D. N., & Simmons, R. (1988). Patterns of misunderstanding: An integrative model for science, math, and programming. *Review of Educational Research, 58*(3), 303–326. https://doi.org/10.3102/00346543058003303

Posner, G. J., Strike, K. A., Hewson, P. W., & Gertzog, W. A. (1982). Accommodation of a scientific conception: Toward a theory of conceptual change. *Science Education, 66*(2), 211–227. https://doi.org/10.1002/sce.3730660207

Rusanen, A.-M., & Pöyhönen, S. (2013). Concepts in change. *Science & Education, 22*(6), 1389–1403. https://doi.org/10.1007/s11191-012-9489-x

Salame, I. I., Krauss, D., & Suleman, S. (2022). Examining learning difficulties and alternative conceptions students face in learning about hybridization in organic chemistry. *International Journal of Chemistry Education Research*, 83–91.

Sari, R. L. P., Pratomo, H., Yunita, I., Purtadi, S., Narayan, M., & Sugiyarto, K. H. (2023). Misconception in chemistry textbooks: A case study on the concept of quantum number, electronic configuration and review for teaching material. *Foundations of Chemistry, 25*(3), 419–437. https://doi.org/10.1007/s10698-023-09475-w

Sawyer, K. R. (2006). The new science of learning. In R. K. Sawyer (Ed.), *The Cambridge handbook of the learning sciences* (Vol. 1, pp. 18).

Schwichow, M., Anna Hellmann, K., & Mikelskis-Seifert, S. (2022). Pre-service teachers' perception of competence, social relatedness, and autonomy in a flipped classroom: Effects on learning to notice student preconceptions. *Journal of Science Teacher Education, 33*(3), 282–302. https://doi.org/10.1080/1046560X.2021.1913332

Slotta, J. D., & H., C. M. T., & Joram, E. (1995). Assessing students' misclassifications of physics concepts: An ontological basis for conceptual change. *Cognition and Instruction, 13*(3), 373–400. https://doi.org/10.1207/s1532690xci1303_2

Smith, E. L., Blakeslee, T. D., & Anderson, C. W. (1993). Teaching strategies associated with conceptual change learning in Science. *Journal of Research in Science Teaching, 30*(2), 111–126. https://doi.org/10.1002/tea.3660300202

Symington, D., & Kirkwood, V. (1996). Lecturer perceptions of student difficulties in a first-year chemistry course. *Journal of Chemical Education, 73*(4), 339. https://doi.org/10.1021/ed073p339

Taber, K. (2002). *Chemical misconceptions: prevention, diagnosis and cure* (Vol. 1). Royal Society of Chemistry.

Taber, K. S. (2001). Building the structural concepts of chemistry: Some considerations from educational research [10.1039/B1RP90014E]. *Chemistry Education Research and Practice, 2*(2), 123–158. https://doi.org/10.1039/B1RP90014E

Taber, K. S. (2003). Mediating mental models of metals: Acknowledging the priority of the learner's prior learning. *Science Education, 87*(5), 732–758. https://doi.org/10.1002/sce.10079

Taber, K. S. (2009). Challenging misconceptions in the chemistry classroom: resources to support teachers. *Educació química,* (4), 13–20.

Taber, K. S. (2013). *Modeling learners and learning in science education.* Springer.

Taber, K. S. (2019). Alternative conceptions and the learning of chemistry. *Israel Journal of Chemistry, 59*(6–7), 450–469. https://doi.org/10.1002/ijch.201800046

Talanquer, V. (2006). *Commonsense chemistry: A model for understanding students' alternative conceptions* [811]. Easton, Pa.

Tan, K.-C.D., Taber, K. S., Goh, N.-K., & Chia, L.-S. (2005). The ionisation energy diagnostic instrument: A two-tier multiple-choice instrument to determine high school students' understanding of ionisation energy. *Chemistry Education Research and Practice, 6*(4), 180–197.

Tanner, K., & Allen, D. (2005). Approaches to biology teaching and learning: Understanding the wrong answers—Teaching toward conceptual change. *Cell Biology Education, 4*(2), 112–117. https://doi.org/10.1187/cbe.05-02-0068

Tokgöz, E. Ö., & Sarıkaya, M. (2025). Experimental study on eliminating misconceptions regarding the law of conservation of mass in chemical reactions in gifted students. *Journal of Chemical Education.* https://doi.org/10.1021/acs.jchemed.4c01285

Tsaparlis, G., Pappa, E. T., & Byers, B. (2018). Teaching and learning chemical bonding: research-based evidence for misconceptions and conceptual difficulties experienced by students in upper secondary schools and the effect of an enriched text [10.1039/C8RP00035B]. *Chemistry Education Research and Practice, 19*(4), 1253–1269. https://doi.org/10.1039/C8RP00035B

van Es, E. A., & Sherin, M. G. (2008). Mathematics teachers' "learning to notice" in the context of a video club. *Teaching and Teacher Education, 24*(2), 244–276. https://doi.org/10.1016/j.tate.2006.11.005

Özmen, H. (2004). Some student misconceptions in chemistry: A literature review of chemical bonding. *Journal of Science Education and Technology, 13*(2), 147–159. https://doi.org/10.1023/B:JOST.0000031255.92943.6d

ÖZmen, H., & Ayas, A. (2003). Students' difficulties in understanding of the conservation of matter in open and closed-system chemical reactions [10.1039/B3RP90017G]. *Chemistry Education Research and Practice, 4*(3), 279–290. https://doi.org/10.1039/B3RP90017G

Models, Their Nature and Role in Chemistry and in Teaching

5

Models, their nature and use in science is a topic that has been largely explored in science education. In this chapter, focus is put on what students should know and learn about models and their use in the teaching–learning process, and on the teachers' teaching and learning responsibilities concerning models and their use in chemistry classrooms.

5.1 What Are Models in the Science Education Context?

First, let us contextualize our use of the word or term "model" to have a clear description and common understanding of what we mean when we use the term. Let us borrow and apply Ludwig Wittgenstein's philosophical conception of a language game and contend that "the meaning of a word is how it is utilized in the language" (David, 2004; Hacker & Schulte, 2009; Kenny, 2006; Wittgenstein, 2009). This could presumably imply that we as a community (scientists, science educators, teachers, and students) possess and retain the power to give meaning to the word or term "model" and hence its function within the context that we have defined ourselves. If we choose to agree to the idea that the meaning of a word or term is contained in how this word or term is used in a given context, then we don't have to worry about a word having different meanings, since we will only focus on that meaning and function that we choose to associate with (attach to) specific words or terms when we use them within contexts we ourselves have established and therefore understand better. Understanding or communication difficulties will only arise if we use the same word or term, for example the word "model" in a communication exchange with others outside of our community, given that they might not share or attach a similar meaning or function to this word. Under these circumstances, it is important to make clear what one implies when they use the

F. Kayima, *Preparing for Chemistry Teaching*, Springer Texts in Education,
https://doi.org/10.1007/978-3-031-97122-8_5

word "model" in a communication context that involves members outside of his/her own community.

With the above understanding in our minds, the best way for us to understand and be able to conceptualize what models are is to consider them in terms of their role and place in science education (teaching and learning) and in the formation and organization of scientific (science) knowledge. On one hand, models are "descriptions of complex systems" (Trindle, 1984), tools used to describe, explain, predict and thereby understand and make sense of natural phenomena (Lazenby et al., 2019). That is to say, every single model is a simplified representation or description of a given scientific phenomenon as a way of facilitating access to, and understanding specific aspects or targets of that phenomenon (Bailer-Jones, 2003; Coll, 2006; Gilbert et al., 1998). On the other hand, models, while essential for the acquisition and organization of scientific knowledge (Frigg & Nguyen, 2017; Koponen, 2007), they themselves form the complete package of the products (outcomes) of scientific work that which includes, science concepts, models, laws, theories and principles (Bergqvist et al., 2013). We could say that the use of models in science is two-fold. First, scientists construct models to represent a given aspect of a phenomenon as a way to provide simplified access to the existence of the observed or explored phenomenon. Second, they then construct and use models to explain target aspects or parts of the observed phenomenon. Thus, we talk of ***a model for a phenomenon or certain aspects or parts of a phenomenon*** and ***a model for explaining the modelled phenomenon***. In other words, science knowledge/scientific facts, once derived, are majorly presented in the form of models and then models are used to describe the modelled concepts.

5.2 Classification of Models Used in Science Education

In many published science education reports (articles, books and book chapters), models are described and classified into different categories based on either their characteristic nature and form (appearance), their ontological status, or based on their subject domain associations (subject domain specific) (see for example; Boulter & Buckley, 2000; Gilbert, 2004; Harrison & Treagust, 2000). Two broad categories of models are generally described. Category one comprises those models that are physical—concrete objects existing separate from the phenomenon they seek to model or represent (pictures, maps, diagrams, etc.). Category two models on the other hand are those models comprising abstraction entities imagined and treated as objects, such as forces, energy, etc. (Gilbert et al., 2000). Others describe models in this category as conceptual/symbolic models—that use symbols and mathematical relationships to describe, predict aspects of phenomenon (Coll, 2006). Regardless of their form and appearance, whether conceptual or mathematical, any given model is constructed to only serve a specific purpose and hence inadequate when the need to explain or describe other aspects of phenomena arises. Moreover, models are only useful to the extent they are able to give a satisfactory explanation of interest about phenomena. They are temporary, bound to change

Table 5.1 Five modes of representations of models: a combined summary of descriptions by Boulter and Buckley (2000) and Gilbert (2004)

Mode of presentation	Description with examples
Concrete (Material)	3D material models, e.g., a plastic heart; a plastic ball-and-stick model of an ion lattice, a coloured plastic model of the human circulatory system, a metal model of an aeroplane
Verbal	Models that are heard or read, of description, explanation, narrative, argument, analogy, and metaphor, e.g. 'The heart is a pump.'
Symbolic	Consists of chemical symbols and formula, chemical equations, and mathematical expressions, particularly equations, e.g., the universal gas law, the reaction rate laws
Visual	Models that are seen, such as graphs, diagrams, animations, some simulations, video, e.g., circle and line drawing of eclipse. Two-dimensional representations of chemical structures (diagrams') fall into this category as do the 'virtual models' produced by computer programmes
Gestural	Models that are movements of the body or its parts, e.g., a solar system made of pupils moving around each other; the representation of the movement of ions during electrolysis by means of pupils moving in counter-flows

over time and are revised or discarded once new understandings and better ways to explain phenomena emerge (Gilbert et al., 2000; Harrison & Treagust, 2000; Justi & Gilbert, 2003; Treagust et al., 2002; Trindle, 1984).

Below are three examples of classifications of models relevant in science education. The first two, respectively presented in Table 5.1 (described for example by Boulter and Buckley (2000) and Gilbert (2004)) and in Table 5.2 (described in published works by Black (1962) and Harrison and Treagust (2000)), are based on appearance (mode of representation and conception), while the one provided in Table 5.3 is based on the ontological status (Gilbert et al., 2000). Understanding the distinction between a classification system based on appearance or form of model (mode of representation) and that based on the ontological status underlying a given model, provides a basis for how best to approach teaching and learning about, and using models in chemistry teaching. I will come back to this important aspect later.

5.3 Understanding the Central Role of Models in Science and Science Education

For purposes of simplicity, we will consider three target groups and discuss the role and use of models with respect to these three, namely (i) the chemists (researchers or scientists), (ii) the teachers (and educators), and (iii) the learners.

Table 5.2 Classification of models by Black (1962) and Harrison and Treagust (2000)

Type	Description
Scale models	Highlight external structure or proportions: model conserves the relative proportions of the original object, and all parts of the model are changed at the same rate An exact copy of an object (e.g. a house, a bridge) on another scale. scale models facilitate looking at something by enlarging it (e.g. a plastic model of a snowflake) or shrinking it (e.g. a globe as a model of the earth)
Analogue models	Models involving change of medium—some material object, system or process designed to reproduce as faithfully as possible in some new medium the structure or web of relationships in the original. Involves the comparison between two things that are similar in some respects
Mathematical models	Models that quantitatively represent relationships between variables, for instance in the form of equations or graphs
Theoretical models	Models that describe the unobservable entities believed to be responsible for natural occurrences

5.3.1 For Chemists (Scientists)—a Model for a Phenomenon and a Model for Explaining the Modelled Phenomenon

Understanding the nature of science itself is a fundamental prerequisite if one is to explore and fully understand the central role and place of models in science. For example, how do scientists generate and communicate scientific knowledge (or facts)? In what form is scientific knowledge once derived, presented (communicated)? Models themselves form part of what constitutes science knowledge, along with concepts, laws and theories (Bergqvist et al., 2013). In the scientists' quest to understand the experienced natural phenomena through scientific investigations, the resulting descriptions, predictions, explanations and hence science knowledge, are in the form of concepts, models, theories, and laws. In this respect, a model is the scientist's medium with which s/he produces and communicates derived understandings to make them visible (wherever applicable), fathomable or comprehendible to the target audience. Rosenblueth and Wiener (1945) described this as ***abstraction***, "replacing the part of the universe under consideration by a model of similar but simpler structure" (p. 316).

Indeed, ***abstraction*** helps researchers to grasp and control certain elements of the phenomena they seek to understand, and this is achieved through developing/suggesting a model for either a part of, or the entire phenomena under investigation. We can conceptualize the process of ***abstraction*** to be occurring at three levels, that is, (i) when a model is used to substitute or represent the explored phenomenon, and (ii), when a model is developed as an explanation for the modelled phenomenon at level (i). The third level of abstraction applies more to teachers (educators) and learners and hence we will address it in the next sections. To exemplify the first two levels of abstraction, let us consider earlier scientific efforts that sought to describe the strange behavior of *absolutely small particles like electrons*. Accordingly, the idea that very small particles like the *electrons of*

Table 5.3 Ontological classification of models in science and science teaching by Gilbert et al. (2000)

Model name	Description
Mental models	Personal, cognitive/ internal representations of external reality by an individual, alone or within a group. They are constructions based on the individual's unique life experiences, perceptions, and understandings of the world. Using mental models, the individual is able to anticipate events, reason, and form explanations (Jones et al., 2011)
Expressed models	When individuals share and discuss their mental models about phenomena with the public, that is, when they express their internal representations, conceptions and constructions that illuminate their understanding of external reality, the resulting models are called *expressed models* The resulting expressed model is not exactly the same as the original mental model. When the individual starts to express his/her mental idea, the process of expressing itself changes the idea and then it is no longer the same original idea
Consensus models	Once the expressed model, having subjected to discussion and or experimentation, is accepted to be of value by the public or by an interest group or community, for example the science education community, then we have a *consensus model* *A scientific model* is *a consensus model* having been accepted by a community of scientists following experimental testing and consequent publication of resulting evidence
Historical models	As explorations continue to be made about a phenomenon of interest, new knowledge, understandings and perspectives emerge and thereby replacing earlier understandings. Consensus models become historical models this way once they are superseded my newer models resulting from improved or further understanding of the explored phenomenon
Curriculum models	Curriculum models are either historical or scientific—consensus models included in a formal curriculum. Curriculum developers and textbook authors often strive to develop simpler representations of these models
Teaching models	Science educators and researchers develop models to assist in the learning process. The goal is to facilitate understanding of scientific concepts, laws, theories, models (historical or consensus) and principles

(continued)

Table 5.3 (continued)

Model name	Description
Hybrid models	A hybrid model is developed for curriculum and classroom teaching purposes. It bears characteristics of at least one of other models. E.g., It can be made by merging some characteristics of the historical/ consensus and curriculum models
A model of pedagogy/ Didactic models	A model of pedagogy or didactic model is developed to serve as a guiding framework to identify and characterize important elements of teaching and learning at general and domain-specific levels. These models guide teachers' decisions and actions regarding why, what, how, and when to teach certain content or implement a specific instructional activity

a given atom, could exist in two different states at the same time was (at the time) and still incomprehensible outside the world of quantum-mechanics. How could something be in two different states simultaneously? Moreover, the revelation that any attempts to measure (and or observe) the particle while in two states, forces the particle into one state or the other was also incomprehensible and seemed like one of trying to avoid the burden of logical reasoning.

Having grappled with this *quantum strangeness and absurdity*, Erwin Schrödinger (1887–1961) devised and described a *thought experiment* termed "Schrödinger's Cat" and used the idea as an abstraction of the rather complex and unfathomable behavior of absolutely small particles. From this example, there were two complex challenges faced by proponents of quantum-mechanics relating to the discovery of the atom, its constituent parts, and characterizing the behavior of the atom's constituents. First, they needed to make the idea of the strange behavior of "absolutely small particles" comprehensible to the science community. At this stage, revelations of what happens at the microscopic level could only be communicated via some kind of representation (a substitute) for the idea to be grasped or qualified. This is the first level of the abstraction process, and at this stage, we talk about a model representing the explored phenomenon, or derived ideas and revelations (facts or knowledge). We can call this packaging derived scientific facts, understandings or ideas into a simpler and more human language and experience related formats. Second, having packaged the ideas, the next stage of abstraction involves breaking down the packaged concepts further for others to understand (a wider community beyond experts in the science field). It is here at this level that we have the "Schrödinger's Cat". It is a simpler way of visualizing the idea that something can exist in two states simultaneously, that is, without access to the internal activity inside the box, the cat is imagined dead and alive at the same time (you can read more about Schrödinger's Cat for more details).

5.3.2 For Teachers (and Educators)—Teaching and Learning

Let us consider three functions of models as generally described in several science education literature, that is, models are, (i) produced as simpler forms or representations of abstract objects, ideas, and concepts, (ii) produced to enable visualization of complex or abstract objects, systems, ideas or concepts, and (iii) produced to provide explanations for complex or abstract systems, ideas or concepts. We have already established from the sections before that the models as understood in the science education context are a part of the products of scientific investigations. We also mentioned that at least two levels of abstraction had been employed by earlier scientists (researchers) as they sought to break down and communicate their research revelations. These two abstraction levels, as explained earlier, result in models representing the explored phenomenon and models that seek to explain or make the modelled phenomenon visible at level one of abstraction. Note that there are instances where the initial model serves at both levels of abstraction, that is, when the model of the explored phenomenon allows understanding the phenomenon without the need for another model to explain the modelled phenomenon. With this understanding, let us now address the teacher and his or her use of models in teaching.

We are all aware that a science teacher's role is to facilitate the learning of science (science concepts, science process and skills and the applications of science) being guided by a given curriculum. Thus, the overall goal of a teacher's use of models in teaching can be thought to be that of making science concepts digestible to the students. However, since a huge part of science knowledge are models themselves, we must think about models in science classrooms in at least three dimensions. First, models as the science to be learned, second, models as representations that meet and align with cognitive limitations and boundaries of the seeker (the person who seeks to understand science concepts, e.g. students), and third, models as tools of explanation, language, and a communication medium. When planning for science teaching and during the actual teaching process, the teacher thus has several options for selecting what models to employ when communicating and explaining the intended science for students to understand. S/he can choose to go directly and use original ideas as they were originally packaged and communicated by their respective pioneers, or s/he could embrace a level three abstraction that involves either developing own but more simplified representations of scientific phenomena. The teacher, however, needs to have a good understanding of the original ideas describing the phenomena (including an understanding of those models which are themselves the science to be taught or learned) to be able to select or develop representations that would serve the intended learning purpose without resulting in students misunderstanding the original idea.

In Fig. 5.1, three levels of abstraction are shown. At level one, the model is used as a packaging of the revealed or derived science ideas/concepts. We can say that at this level, the model itself is the product or outcome of the performed scientific work (scientific investigation). At level two, the model is used to make the revealed idea fathomable, visible, etc. Models at level two abstraction serve

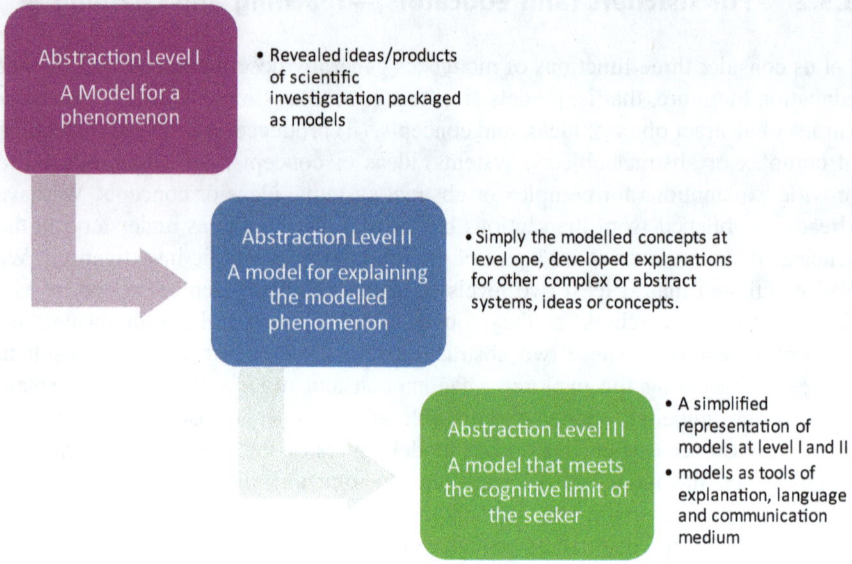

Fig. 5.1 Levels of abstraction of revealed or derived science ideas and concepts

either to simply the modelled concepts at level one, or they could be developed explanations for other complex or abstract systems, ideas, or concepts. Quite often at this level two of abstraction, researchers (investigators) strive to develop models that are thought to meet and align with cognitive limitations and boundaries of a wider community who are the target consumers of science. At level three, the developed model also serves to match the contextual level of those that seek to understand the original idea presented at either level one or two. However, this level has more degrees of freedom and model developers have an open-ended space and opportunities to suggest and use whatever forms of expressions to facilitate access to and understanding of the science modelled at levels one and two. Indeed, it is at level three of abstraction where we have a diversification of model ideas, forms, and topologies, ranging from science leaning and subject specific models to curricula and didactic models. At level three, the overall aim of a model is to facilitate understanding and hence models occur in various and diversified forms, as real (concrete), imagined and theoretical frameworks and tools for explanation, language and communication.

5.3.3 For Learners—Models as Science Knowledge and as Study Tools and Visualization Frameworks

Being students and whose main goal (interest) is to learn and be able to apply science knowledge and skills, models serve majorly two functions (purposes). First, we already mentioned that a good part of science knowledge is presented as models and in this respect, such models are considered the target science to be learned by the students. Second, when models are not the science itself to be learned, they are tools and frameworks used to enable visualization of the would be very complex or abstract systems, ideas, and concepts. Like teachers, students also have the opportunity to navigate through all the three levels of abstraction shown in Fig. 5.1. The difference here is that the teacher has a central role in facilitating students as they navigate through these levels. Moreover, what we have not concretely mentioned so far is that the development of models as frameworks to facilitate or illuminate one's understanding at level two and three (Fig. 5.1) is open to all, including students. For example, at level three, students can develop their own interpretive frames that resonate and align with their own experiences and within their cognitive boundaries and then use these as departure points for understanding the central scientific ideas. In other words, students can and are able to carry out their own abstraction processes of the already abstracted phenomena in order to achieve a more simplified model that is closer to their own individual experiences.

Consequently, modelling and development of models, and the thinking process that leads to derived understandings are intertwined. What happens during cognitive processing is indeed modelling. Indeed, derived understanding only occurs after an individual successfully develops some kind of model—a packaging to represent that understanding. The products from this cognitive processing effort is what has been conceived as mental models in literature as I will elaborate later. It is important thus to recognize that the process of abstraction is not exclusive to only researchers and teachers, but it is for all, including students. With this understanding, you realize that developing models as tools with which to visualize and understand science concepts is a continuous process which is part of the students' cognitive processing that leads to derived understanding. The teacher on other hand can do the same, that is, develop models (initially mentally) and use these to facilitate students' access and understanding of abstract concepts.

5.4 Scientific Models: A Product of the Science Enterprise, a Science to Be Learned, and a Tool to Aid Understanding

The discussion so far has been more focused on models in term of their inherent role or function as simpler forms or representations of abstract systems—objects, ideas, and concepts, and as tools that explain and enable visualization of complex systems. We will now closely look at the place of scientific models as a part of science knowledge. The aim is to emphasize that teaching of and about models should be considered along these two dimensions, (i) tools that facilitate understanding, visualization of abstract systems, and (ii) as knowledge to be learned/taught as one of the products of scientific endeavors.

To enable visualization of scientific models as one of the products of the scientific research endeavor, I will consider a brief description of the research process that results in what counts as science knowledge (scientific knowledge). As our departure point, let us consider a summary in Fig. 5.2 of Chalmers' (2013) careful elucidation of the common view that scientific knowledge is based on facts established through careful and unprejudiced observation and experiment. The process can be thought to go through at least four stages which are subject to several iterations (back and forth) at the discretion of the researcher. First, it starts with the questions—the arising or existing problem—an issue that requires an answer, or a mind in a state of wonder—seeking to make sense of a certain phenomenon and hence setting a set of research questions or what often is termed as "the hypothesis" to be tested, confirmed, or falsified. This first stage is both informed by and guided by the researcher's knowledge background—"theoretical commitments, beliefs, previous knowledge, training experiences, and expectations…." (Lederman, 2007, p. 834). We can call this a person's personal interpretive framework - a set of cognitions, abilities, personal experiences, training experiences, and expectations (Kayima, 2022). This personal interpretive framework is also informed by and shaped by the individual's social and cultural context.

Once the researcher has decided upon what he/she wishes/intends to explore, the next and second stage is to establish how s/he will explore and find answers to the pending question or problem. This is where we have the question of the *scientific method* that will lead to objective science facts. The reference to "scientific method" and not "methods" is not to imply an existence of a single well-defined method or approach with which to explore and find answers to the posed research problems. It is rather an emphasis on the acceptable research practice that consists in following a certain set of guidelines to ensure objectivity, trustworthiness, and the extent to which it is possible to employ the research in different contexts. Commonly known approaches for gathering data that informs the explored problem as summarized in Fig. 5.2, involve making observations (what we can see, hear or touch) and experimentation. Regarding observation, the expectation is that these will be carried out in a careful and unprejudiced way as a prerequisite for a secure and objective basis for science (Chalmers, 2013).

At stage three, the researcher has amassed sufficient data through either laboratory experimental work, recordings and transcriptions of observations, or any other form of legally generated data. It is these data sets that will form the factual basis for what would be the resulting science (scientific) knowledge. However, to derive or draw any inferences from these data sets, the researcher must engage the data with his/her own interpretive framework. The function played by one's interpretive framework is independent of whether the researcher is working with a pre-established reference standard, procedure, a guiding analytical framework or not. The researcher does hence have to invoke what Chalmers termed as "a careful and unprejudiced use of senses" when engaging data to derive science facts. Moreover, mentioning the "use of senses" does not only imply a limitation to the common five senses. The statement includes an implicit inclusion of the cognition dimension, and it implies engaging data with an entire set of the researcher's

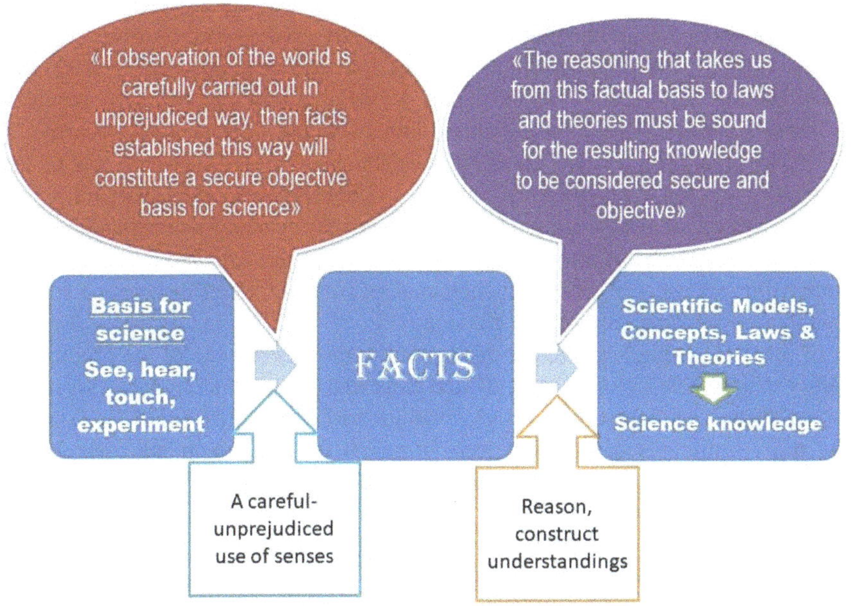

Fig. 5.2 Developed as a summary for the process resulting in scientific knowledge as elucidated by Chalmers (2013, p. 1)

interpretive framework. When the researcher succeeds in his/her engagement with data and manages to derive some facts, the next and fourth stage is to make sense of these derived facts. Chalmers notes that (see Fig. 5.2) "the reasoning that takes the researcher from the derived facts to generating models, theories and laws, "must be sound for the resulting knowledge to be considered secure and objective" (Chalmers, 2013, p. 1). In other words, the researcher must again engage with the derived facts with his/her interpretive framework to construct a logical, consistent, and representative understanding of the derived facts.

From the overview above, what really happens in the process of constructing understandings from collected data (evidence), and when reasoning from derived facts towards onto the resulting scientific knowledge, is modelling and model construction. Models and modelling are not simply simplifications of complex concepts or systems, it is our way of capturing, qualifying, and packaging our understandings of the explored part of the world and when expressing or sharing these understandings to others. We cannot imagine ourselves separated from this, since it is what defines us. Our understanding of the world are indeed our mental constructions influenced by our individual interpretive frameworks, which we later express in different external forms and representations. Any model, regardless of its genre, characterization, or form of appearance, starts as mental constructions depicting and capturing the way the individual understands or conceptualizes the explored problem. These constructions emerge because of the enacted engagement between the individual's personal interpretive framework and the available evidence (data), based upon which constructions are developed. It is these constructions that are referred to as mental models. As in Fig. 5.2, the result of the sense making process (engagement with collected evidence/data to arrive at science facts, and then engagement with established facts to arrive at new understandings—scientific knowledge) could be (i) developed models that illuminate, explain, and or otherwise simplify and represent the revealed understandings, (ii) laws—which are descriptions of or statements about the revealed and established relationships between explored phenomena, or (iii) suggested theories to explain the explored phenomenon or its behavior.

To illustrate further concretely that scientific models are one of the products of science research in as much as they are tools that simplify and enable visualization of abstract systems/ideas, let us consider earlier research efforts to establish how the constituents of an atom were arranged in an atom in the early 1900s. By the time, the following information had been fully established, that's, (i) the electron had been discovered (by J.J Thomson in 1897) and confirmed to be negatively charged, (ii) the atom had been established to carry neutral charge, and (iii) the supposed existence of an equally positively charged particles was predictable to counteract the negative charge and ensure a neutral atom. The question that needed to be answered at the time was thus about how the very small particles were arranged in the atom. As the process illustrated in Fig. 5.2 (that is, through imagination, observation, and experimentation), the exploration of this question resulted into the development and advancement of at least four consecutive kinds of models of how the constituents an atom were arranged within the atom, namely (a) the **"plum pudding"** model (by J.J Thompson in 1904), (b) the **nuclear model** (by Rutherford and colleagues around 1911), (c) **The Bohr-Rutherford atomic model** (by Niels Bohr and Ernest Rutherford in 1913) and (d) The **Quantum mechanical model of the atom**—1926 (Erwin Schrödinger/Werner Heisenberg). Table 5.4 shows a summary of the central ideas for each of the proposed or derived model and accompanying pictorial representations of these central ideas respectively.

Looking at the summaries in Table 5.4, the central idea revolves around the revealed understandings of how the constituents of the atom are organized, and the descriptions provided in the middle column are indeed the available revealed understandings at different given timelines. For example, Thompson's Plum pudding model of the atom before it was proved wrong by his student Ernest Rutherford and subsequently replaced by nuclear model and later the Bohr-Rutherford atomic model, was once the understanding revealed by the interest investigator (Thompson) at the time. Note that as discussed in Chapter 1, science knowledge is not a finished product, and scientific discoveries are continuously emerging as we humans are continuously exploring natural phenomena, adding to existing understandings, while refuting and correcting existing wrong understandings once new or better understandings emerge. This truism is clearly visible as you move down the timeline in Table 5.4. That said, there are two things to grasp here. First, apart from the revealed descriptions described about the how the

Table 5.4 The Road to understanding the structure of the atom

Model	Central idea (s)	Figure/illustration
(a) the **"plum pudding"** model (by J.J Thompson in 1904)	The positive charge is continuously distributed throughout the atom negatively charged electrons are distributed in the atom like raisins inside a cake or like plums in a pudding, …." (Hentschel, 2009, pp. 18–19)	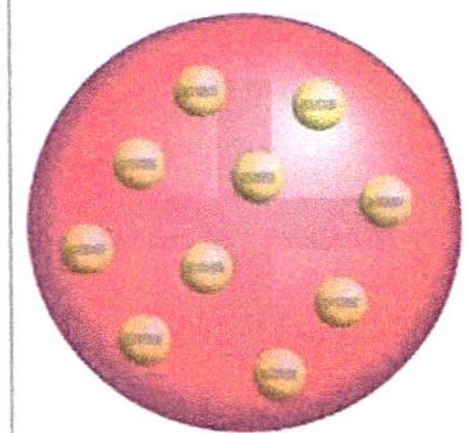 Credit: User: Fastfission/Wikimedia Commons; Source: http://commons.wikimedia.org/wiki/File: Plum_pudding_atom.svg(opens in new window); License: Public Domain)
(b) the **nuclear model** (by Rutherford and colleagues around 1911)	The electrons revolve in empty space around the nucleus that is positively charged and contains most of the atomic mass (Podgoršak, 2006) There are many negatively charged electrons outside the nucleus as there are positively charged particles within the nucleus—hence an electrically neutral atom." (Nivaldo, 2017, p. 92)	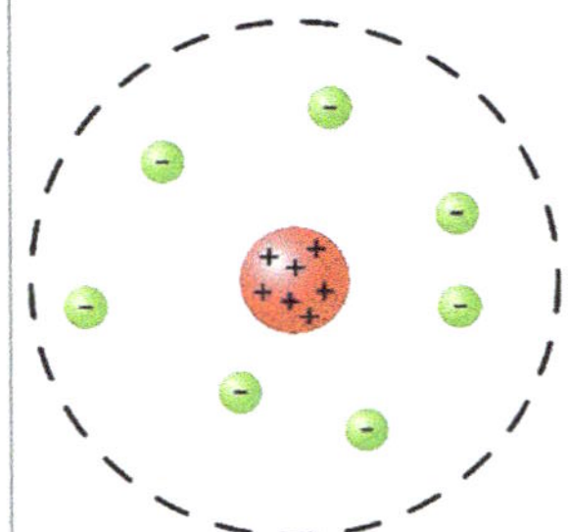 By Bensteele1995 - Own work, CC BY-SA 3.0, https://commons.wikimedia.org/w/index.php?curid=21332267

(continued)

Table 5.4 (continued)

Model	Central idea (s)	Figure/illustration
(c) **The Bohr-Rutherford atomic model** (by Bohr and Rutherford in 1913)	(i)The atom is mostly empty space with a massive and dense nucleus with electrons in the center. (ii) Electrons orbit the atom's nucleus in orbitals or energy levels (iii) Each orbital can have more than one electron (iv) An electrically neutral atom has as many electrons as there are protons	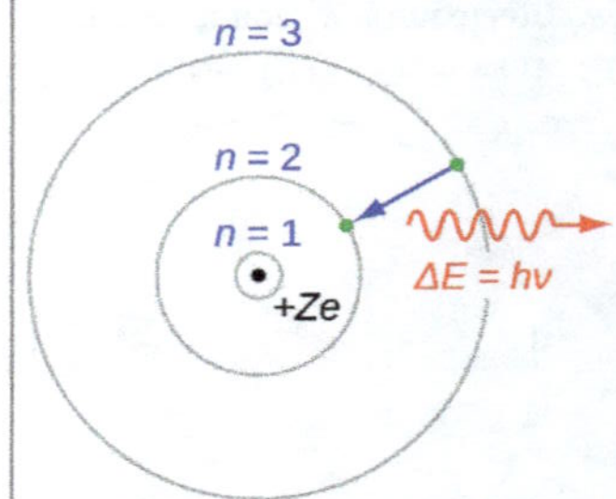 By JabberWok, CC BY-SA 3.0, https://commons.wikimedia.org/w/index.php?curid=2639910
(d) The Quantum mechanical model of the atom—1926 (Erwin Schrödinger/Werner Heisenberg)	(i) Electrons exist in a cloud around the nucleus in areas called orbitals and do not exist in fixed orbits (ii) Electrons exist in two states at the same time, e.g., they are particles and waves at the same time (iii) The location of electrons cannot be pinpointed, rather it is described using probability. Orbitals are thus areas of high probability of finding the electron	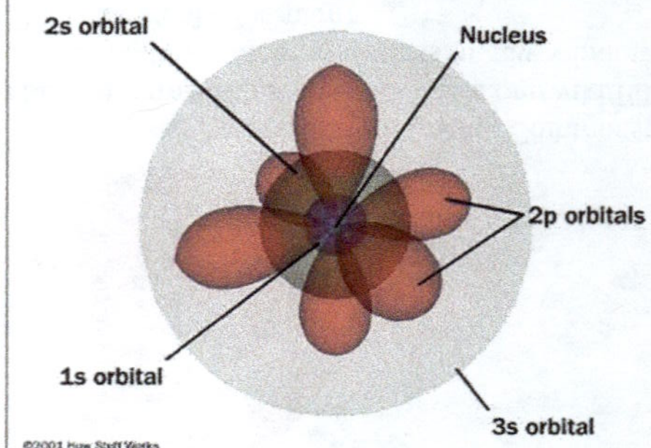 Source: Craig Freudenrich, Ph.D. "How Atoms Work" 1 January 1970. HowStuffWorks.com. < https://science.howstuffworks.com/atom.htm > 7 November 2024

constituents of the atom were organized at a given timeline (let us say the plum pudding model—Thomson 1904, the nuclear model—Ernest Rutherford 1911, the Planetary model—Niels Bohr 1913), we do not have any other access and neither a different revealed and accepted understanding at any given specific timeline, not until new understandings from the continuing scientific exploration emerge. Until proven otherwise, what has been revealed and accepted by the scientific community at any given timeline serves as the science knowledge at that time. Since the descriptions in middle of Table 5.4 are what we have as the revealed understandings and accepted descriptions of how the constituents of the atom are arranged at given historical points in the history of the discovery of the atomic structure, this information is not merely a model of how the electrons and protons are organized within an atom, but is the only knowledge we have about the structural organization of the constituents of the atom, and hence the science knowledge about the structure of the atom at any one given timeline (at a given historical point). Therefore, the different models of the atoms described at the different timelines serve as both models of the atom structure and as the science knowledge about the atom structure. The figures presented in the third column respectively are illustrations of the information provided in the middle column and as I showed in Fig. 5.1 earlier, they are a form of level three abstraction, to make it easier to visualize and grasp the ideas presented in the middle column.

5.5 What Should Be Taught to Learners About Models?

In the above section, we have gone through some conceptual understandings to help us understand the place and role of models in science and in teaching. We will now use this as our background knowledge as we consider the question of *what learners need to learn about models and why*. First, I want to assume that there are not very many chemistry classroom teachers in schools who directly take up and teach "models and their use" as one of the topics in their respective teaching. This is in part because "models in science" is not often considered in most school curricula as a topic to be taught even though teachers use models all the time in their teaching as tools of communication, for visualization of abstract concepts, and in cases where the models themselves are the science to be learned. The other reason relates to teachers being unaware of what to teach about models and how to teach it even though the curriculum provides the opportunity for one to teach about models as part of their teaching. On the other hand, in some countries, modelling and the use of models is a central element in the science curriculum and students are expected to learn about, understand and be able to use models to advance their science understanding. Regardless of whether "models in science" is considered a topic in curriculum or not, we all agree that both teachers and students use models to talk about, discuss and understand science, and hence understanding and being able to use models in science teaching and learning is both important to teachers and students.

Moreover, the picture is slightly different when we consider what happens during teacher training at College and University level. Models and their use in science and in teaching is one of the science didactics topics considered and taught in science education at teacher education colleges and universities. In most cases, when educators introduce the use of models to their students (teacher students/trainees), emphasis is put on describing models as representations of science phenomenon (an object, system, or process), as anything known (a photo or pattern, etc.) used to describe something new and unknown. This is proceeded by describing the different types of models as well as the strength and weaknesses or shortcomings of models used in science. When educators discuss the different types of models in science, the most frequently embraced classification system is that which distinguishes models based on appearance or mode of representation as earlier shown in Tables 5.1 and 5.2, where models are characterized as concrete, verbal, symbolic, visual, scale, gestural analogic, theoretical, mathematical, or graphical, etc.

The problem of approaching the teaching of and about models by classifying them based on the mode of presentation or form is that it denies students the ontological understanding of the place and role of models in the development of scientific knowledge and the opportunity to develop a coherent understanding of the nature of scientific knowledge. If we simply present models to our students using merely physically fathomable classifications, along with a discussion of the limitations and strengths of models, we deny them the opportunity to develop a deeper and transferrable understanding of the relationship between modelling and cognitive processing. It implies that we only focus on models as the end products characterized by what is observable or quantifiable, but not the inherent process underlying the development of what qualifies as a model. This superficial teaching about models does not lead to students understanding what models are, how they come into being, and how the application of models (modelling and use of developed models) is an action intertwined with cognitive processing. W*hat then would be the best way to approach the teaching of models to students?* Let us consider this question in the next section.

5.5.1 A More Effective Way to Approach the Teaching of and Learning About Models in Science

At the beginning of this chapter, I described two kinds of classification of models in science education, namely, one based on the mode of presentation, appearance, or form, and the other based on the ontological status of the model. These two classification categories are shown in Tables 5.1 and 5.2 (Black, 1962; Boulter & Buckley, 2000; Gilbert, 2004; Harrison & Treagust, 2000) for a classification based on mode of presentation and Table 5.3 (Gilbert et al., 2000) for a classification based on ontological status. Please note that there are other category systems for models suggested in different science education literature, however, most of these (if not all) fall into the two category systems mentioned above. A classification of models based on ontological status seems to be the most suitable category system with which teachers can introduce models to their students before going further to classify models based on their appearance or mode of presentation. I previously noted that understanding models for students demands an understanding of the inherent process underlying the development and qualification of what counts as models, that is, understanding how models come into being before describing them based on form, appearance, or mode of presentation. A classification based on ontological status opens the door to this kind of understanding. The ontological classification of models as suggested by among others, Gilbert et al. (2000), mirrors an understanding of the inherent process involved in deriving scientific inferences that result in derived scientific knowledge.

If students are to grasp the central idea about models and be able to identify and use different kinds of models, and start to appreciate, improve, and use their own constructions (models), the most effective way to achieve this is to follow a teaching that does not focus on classifying models based on their appearance or form, but that which starts by seeking to understand how models come into being. This is consistent with understanding models and modelling as an inherent part of what we do every day in our daily lives as we continuously explore and make sense of the world around us. In terms of teaching, it implies starting with addressing how we move from emerging (emerged) mental constructions (mental models) in our heads (minds) to expressing and sharing these constructions or understandings with others (expressed models), and so on. The focus here is not on how the model looks like or appears externally, but on how it is constructed at the beginning and then it goes through a legitimization process, from being a mental model—to the expressed model—to consensus model—and then to a scientific model that's acceptable as part of science knowledge. Other types of models such as curriculum, teaching and pedagogical models also start as mental constructions in the mind of the developer (e.g., the science educator, teacher or researcher) before being expressed and accepted by the wider community for the purposes they are intended to serve.

Moreover, even before going over to discuss models that aid or facilitate the understanding of concepts, such as the Molymod molecular model sets (https://molymod.com/sets/) for visualizing how molecules are structured, the teacher addressing the topic of "use of models in chemistry classrooms", needs to first select and discuss one or two examples of the common scientific models or historical models (models that have been superseded) for example the Bohr model of the atom, Boyles acid–base model, Arrhenius acid–base model, etc. This would enhance students' understanding of scientific models as a part of the science to be learned, as well as the inherent distinction between models that comprise the science knowledge itself and models constructed to aid in the teaching and learning of science. Moreover, it would foster understanding if the teacher were able to explain the argument that what counts as science knowledge also comprises models. Selecting and discussing one or two scientific/historical model examples can start with a brief description of the process involved in generating scientific knowledge, to clearly show students the place of scientific models as a part of the science knowledge itself. This should not be difficult for a teacher who has knowledge about the nature of science.

By considering and critically internalizing the process through which science or scientific knowledge comes into being, the teacher has a clear picture of the existing distinction between the place, nature, and functions of the different types of models that we have or talk about in science education generally. A distinction exists between scientific models which are a part of the science knowledge students are supposed to learn, and other models whose aim is to abstract and represent science concepts (science knowledge generally) in more simpler and fathomable forms to make it easier for those seeking to understand these concepts. As a science teacher, you will need to and should be able to visualize and understand this distinction right from your understanding of the nature of science, what counts as science knowledge and how that knowledge comes into being (scientific methods). This will form a good basis for proper planning of what to start with and how to start when introducing "models and their use" in classroom lessons.

Finally, in science education literature, we have several suggestions regarding what should be emphasized when it comes to students' learning about models in science. Among these is the strong emphasis on students understanding the nature of models, their functions, scope, and limitations of different scientific models. The approach to introducing and teaching models described above enables taking up these suggested elements in moreover a more logical and coherent way that ensures productive understanding. As a prerequisite to this way of approaching the teaching and learning about models and their use in science, the teacher should possess a clear and comprehensive understanding of the nature of models and how these come into being regardless of their genre, topology, and function (whether scientific or didactical models). He/she should be able to demonstrate with an example the different developmental stages of a given scientific model or any other curricula or didactic model, that is, starting as a mental idea or set of mental ideas (mental model) that is expressed either verbally, by symbols, objects, drawings, signs, etc., and then accepted by a wider fellowship and thereby transiting into a consensus model, which later could qualify to be called a scientific model based on which community of people are involved. This process is easy to demonstrate and gives the chance to students to practice their own developing and expression of self-generated models. Moreover, using commonly known science knowledge development transitions like the history of the development of the atomic structure (see Table 5.4), the teacher can show and demonstrate the conception of scientific models superseding one another and resulting in historical models.

References

Bailer-Jones, D. M. (2003). Models, theories and phenomena. *Proceedings of logic methodology and philosophy of science.*

Bergqvist, A., Drechsler, M., De Jong, O., & Rundgren, S.-N. C. (2013). Representations of chemical bonding models in school textbooks—help or hindrance for understanding? [10.1039/C3RP20159G]. *Chemistry Education Research and Practice, 14*(4), 589–606. https://doi.org/10.1039/C3RP20159G

Black, M. (1962). *Models and metaphors: Studies in language and philosophy*. Cornell University Press.

Boulter, C. J., & Buckley, B. C. (2000). Constructing a typology of models for science education. In J. K. Gilbert & C. J. Boulter (Eds.), *Developing models in science education* (pp. 41–57). Springer Netherlands. https://doi.org/10.1007/978-94-010-0876-1_3

Chalmers, A. (2013). *What is this thing called science?* (4 ed.). McGraw-Hill Education.

Coll, R. K. (2006). The role of models, mental models and analogies in chemistry teaching. In P. J. Aubusson, A. G. Harrison, & S. M. Ritchie (Eds.), *Metaphor and analogy in science education* (pp. 65–77). Springer Netherlands. https://doi.org/10.1007/1-4020-3830-5_6

David, G. S. (2004). *Wittgenstein's philosophical investigations : An introduction* [Book]. Cambridge University Press. https://search.ebscohost.com/login.aspx?direct=true&db=nlebk&AN=304498&authtype=sso&custid=ntnu&site=ehost-live

Frigg, R., & Nguyen, J. (2017). Models and representation. In L. Magnani & T. Bertolotti (Eds.), *Springer handbook of model-based science* (pp. 49–102). Springer International Publishing. https://doi.org/10.1007/978-3-319-30526-4_3

Gilbert, J. K. (2004). Models and modelling: Routes to more authentic science education. *International Journal of Science and Mathematics Education, 2*(2), 115–130. https://doi.org/10.1007/s10763-004-3186-4

Gilbert, J. K., Boulter, C., & Rutherford, M. (1998). Models in explanations, Part 1: Horses for courses? *International Journal of Science Education, 20*(1), 83–97. https://doi.org/10.1080/0950069980200106

Gilbert, J. K., Boulter, C. J., & Elmer, R. (2000). Positioning models in science education and in design and technology education. In J. K. Gilbert & C. J. Boulter (Eds.), *Developing models in science education* (pp. 3–17). Springer Netherlands. https://doi.org/10.1007/978-94-010-0876-1_1

Hacker, P. M. S., & Schulte, J. (2009). *Philosophical investigations*. Wiley-Blackwell.

Harrison, A. G., & Treagust, D. F. (2000). A typology of school science models. *International Journal of Science Education, 22*(9), 1011–1026. https://doi.org/10.1080/095006900416884

Hentschel, K. (2009). Atomic models, J.J. Thomson's "Plum Pudding" model. In D. Greenberger, K. Hentschel, & F. Weinert (Eds.), *Compendium of quantum physics* (pp. 18–21). Springer Berlin Heidelberg. https://doi.org/10.1007/978-3-540-70626-7_9

Jones, N. A., Ross, H., Lynam, T., Perez, P., & Leitch, A. (2011). Mental models an interdisciplinary synthesis of theory and methods. *Ecology and Society, 16*(1). http://www.jstor.org/stable/26268859

Justi, R., & Gilbert, J. (2003). Models and modelling in chemical education. In J. K. Gilbert, O. De Jong, R. Justi, D. F. Treagust, & J. H. Van Driel (Eds.), *Chemical education: Towards research-based practice* (pp. 47–68). Springer Netherlands. https://doi.org/10.1007/0-306-47977-X_3

Kayima, F. (2022). The role of reflective practice in mediating development of pre-service science teachers' professional and classroom knowledge. *Interdisciplinary Journal of Environmental and Science Education, 18*(1), e2262. https://doi.org/10.21601/ijese/11364

Kenny, A. (2006). *Wittgenstein* (Rev. ed.). Blackwell.

Koponen, I. T. (2007). Models and modelling in physics education: A critical re-analysis of philosophical underpinnings and suggestions for revisions. *Science & Education, 16*(7), 751–773. https://doi.org/10.1007/s11191-006-9000-7

Lazenby, K., Rupp, C. A., Brandriet, A., Mauger-Sonnek, K., & Becker, N. M. (2019). Undergraduate chemistry students' conceptualization of models in general chemistry. *Journal of Chemical Education, 96*(3), 455–468. https://doi.org/10.1021/acs.jchemed.8b00813

Lederman, N. G. (2007). Nature of science: Past, present, and future. In S. K. Abell & N. G. Lederman (Eds.), *Handbook of research on science education* (pp. 831–879). Routledge.

Nivaldo, T. J. (2017). *Chemistry: A molecular approach* (4th (Global). ed.). Pearson.

Podgoršak, E. B. (2006). Rutherford—Bohr atomic model. In E. B. Podgoršak (Ed.), *Radiation physics for medical physicists* (pp. 42–85). Springer Berlin Heidelberg. https://doi.org/10.1007/3-540-29471-6_2

Rosenblueth, A., & Wiener, N. (1945). *The role of models in science* [316–321]. East Lansing, Mich.

Treagust, D. F., Gail, C., & Mamiala, T. L. (2002). Students' understanding of the role of scientific models in learning science. *International Journal of Science Education, 24*(4), 357–368. https://doi.org/10.1080/09500690110066485

Trindle, C. (1984). The hierarchy of models in chemistry. *Croatica Chemica Acta, 57*(6), 1231–1245.

Wittgenstein, L. (2009). *Philosophical investigations: The german text with an English translation.* Wiley Blackwell.

Assessment in Chemistry 6

Abstract

In the educational context, any act of teaching and learning goes hand in hand with assessment, either to support the learning process or to help establish the extent of attainment of target goals of the teaching–learning program or both. This chapter covers the theoretical perspectives underpinning classroom assessment practices, frameworks and assessment models developed as potential guides to planning and implementing productive assessment, as well as examples of assessment tools (forms, methods, strategies or instruments) relevant to chemistry teaching and learning contexts.

6.1 Classroom Assessment: What Is It About and Why?

In the educational context, any act of teaching and learning goes hand in hand with assessment, either to support the learning process or to help establish the extent of attainment of target goals of the teaching–learning program or both. As such, classroom assessment is distinguished mainly into (i) assessment for learning—often referred to as "formative assessment" and (ii) assessment of learning—often referred to as "summative assessment". Assessment for learning aims to support students' learning. Different teachers, depending on their level of knowledge and expertise, employ varied methods, approaches, techniques, activities, e.tc., to establish where their students are in the learning process, provide feedback to students regarding how to move from where they are to where the teacher wishes them to be, and to use the assessment evidence to adapt the teaching and learning ways that meet students' learning needs as the learning progresses. In other words, assessment for learning serves both the teacher and students in that, the student gets to know what he/she can do and what aspects need to be improved, while the teacher, in addition to providing feedback to the students, s/he uses the assessment evidence to make learning more effective, including making necessary changes to his or her

F. Kayima, *Preparing for Chemistry Teaching*, Springer Texts in Education,
https://doi.org/10.1007/978-3-031-97122-8_6

teaching. Moreover, the process of assessing students for learning can transition into learning itself. This often happens if the assessment process also provides opportunities to students to reflect on their own learning process and develop an awareness of how they come to learn and draw on such awareness to assume a bigger responsibility for their own learning process. Under such circumstances, depending on the predetermined purpose of the assessment, it can be reasonable to divide formative assessment into assessment for learning and assessment as learning.

Assessment of learning on the other hand, serves to establish whether students have attained the intended learning outcomes of the learning program at different pre-planned stages of instruction, either at the end of term, semester, year, etc. It is used to summarize how much learning has taken place compared to the total learning that was set out to be gained. Assessment of learning thus is used to collect and provide evidence of achievement to all concerned interest groups and stakeholders namely, students, teachers and parents, the schools and the wider community. Different countries have different educational structural organizations. In some countries, students must pass the end of year exams to be promoted from one class grade to another. In this case, summative assessments at the end of each term or year are greatly emphasized. There is also increasing reliance on standardized summative assessments, where different countries, either as a block or individually, carry out summative assessments of a given group of year-students with the goal of using the results to make necessary educational reforms. Examples include the PISA[1]—tests—programme for international student assessment (PISA) which tests skills and knowledge of 15-year-old students in reading, mathematics and science, and the TIMSS[2]—Trends in International Mathematics and Science Study, a standardized survey that measures students' competence in mathematics and science in class grades 4 and 8.

Considering assessment at the classroom level, there is a reasonable overlap between assessment for learning and assessment of learning. Carefully planned and executed summative assessment can also support students' learning in as much as a carefully crafted assessment for learning could inform on the student's mastery of the intended learning and hence a form of summative assessment. Assessment as a process of gathering information about what students know, can and are able to do, as well as what they are learning to do, is the responsibility of the teacher, who, while planning the assessment strategies, should clearly work out the intended purpose of each planned assessment. An awareness that there is always the secondary purpose that could be accomplished in addition to the primary purpose of any form of assessment whether formative or summative (Bennett, 2011), grants the teacher a wider working space to plan and use assessment strategies that serve both to support students' learning, while also informing on students' attainment of the target learning outcomes of the entire study package (programme).

---

[1] https://www.oecd.org/en/about/programmes/pisa.html.

[2] https://timssandpirls.bc.edu/timss-landing.html.

6.2 Assessment in Chemistry Classrooms

Understanding assessment in chemistry starts with an understanding of what students should learn, what should be assessed and for what purpose. The answer to what should be assessed in chemistry classrooms is tied to the chemistry curriculum of any given educational context, stretching from the national school curriculum of respective countries and down to the school level curricula. The goals and reasons for chemistry inclusion as a subject in different curricula across different countries are always well stated in those respective curricula. These goals, however, are closely related to and if not similar to the main goals of learning chemistry and the importance of chemistry education to the global society. As such, the goals of assessment in chemistry, regardless of the variations in chemistry curricula structural organization, will often converge and be linked to what students need to study and know about chemistry and why. Chemistry is about "the study of the composition and properties of matter and how and why it undergoes change." (Gibney, 2018 p. 31). The fundamental place of chemistry to our world cannot be overemphasized. Basic needs such as food, clothing, shelter, energy, clean air, water, soil and our general health, are all impacted by chemical knowledge and chemical technologies. The benefits of chemistry thus stretch from knowledge that contributes to improving our way of life, to protecting the planet, and to continuously exploring and understanding our world better. To exploit these benefits, the world will continue to have the need for among others, chemistry teachers, analytical chemists, chemical engineers, forensic scientists, laboratory technicians, pharmacists and pharmacy technicians, quality control chemists, toxicologists, environmental consultants, biotechnologists, nanotechnologists, to mention but a few. Consequently, assessment of students' learning and achievement in chemistry is about assessing their knowledge and skills in chemistry, which include,

- Understanding of chemical concepts, facts, patterns, principles, terminology and conventions
- Acquired laboratory knowledge and skills—the ability to use or manipulate apparatus, to handle materials and perform experiments safely and make accurate observations.
- Understanding of the methods used in chemical investigations and the nature of scientific investigations.
- Knowledge and skills of analyzing and interpreting data from various sources and drawing relevant conclusions.
- Mathematical and computational knowledge and skills—ability to manipulate and translate chemical data and perform calculations.
- Applying chemical knowledge to explain observations and solve problems which may involve unfamiliar situations.
- Selecting and organizing scientific information from appropriate sources and communicating this information in an appropriate and logical manner.
- Understanding and evaluating the social, economic, environmental and technological implications of the applications of chemistry; and

– Ability to make decisions based on the examination of evidence and arguments.

The list above is a general list of the knowledge and skills that could be targeted as assessment goals in chemistry teaching and learning. However, as mentioned earlier, every individual country has defined the target learning outcomes of any chemistry teaching that is included in their respective curriculum. The specific goals of assessment should derive from these started target learning outcomes. For example, Table 6.1 shows some of the target competence aims (goals) for the second-year high school chemistry extracted from the Norwegian school curriculum (as of July 2024), a translation from Norwegian to English. The curriculum, as shown in the table, also provides some hints to the teacher regarding the aspects to consider when assessing to improve learning (formative assessment) and when assessing students' learning (summative assessment).

Considering the information in Table 6.1, a clear relationship between the target competence goals and the nature and purposes for both forms of assessment is visible. Moreover, teachers are provided with the kind of indicators of learners' progress and demonstration of attained competences as a basis to enable teachers in designing and selecting appropriate assessment tools. Taking a leaf from Table 6.1 above, consider (study) the curriculum document in your country to get to an orientation into how the chemistry contents to be learned are aligned with the expected learning outcomes, and the rationales that underpin the promoted assessment processes (or practice).

6.3 Theories and Frameworks that Inform Different Assessment Designs and Models

6.3.1 Theoretical Perspectives Underlying Educational Assessment

In the previous section, it was mentioned that a prerequisite to understanding assessment in classrooms lies in getting to understand what students should learn or what we want our students to learn. This leads us to the concept of educational achievement which entails what should count as the target outcomes of any planned and executed learning program. Historically, the way how teaching and learning, as well as assessment practices have proceeded over the years has been influenced by the broad conception of educational achievement. Cole (1990), describes two conceptions of educational achievement, one characterized by achievement of basic skills and facts, and the other characterized by achievement of higher order skills—*such as critical thinking and problem solving* and advanced knowledge—*such as understanding and expertise.* Accordingly, the former has its roots in behavioral psychology where learning is characteristic of specific, discrete skills with very clear and precise desired outcomes, while the latter has its tradition routed in philosophy and cognitive psychology. Biggs (1995), building upon Cole's (1990) syntheses, distinguished two educational thinking

Table 6.1 Extract of some of the competence goals for chemistry 1 in the Norwegian curriculum for the second-year high school students (as of July 2024)

Target competence goals for chemistry 1: Second year of high school chemistry in Norway: The aim of the training is for the student to be able to;	Purpose for formative assessment—assessment for learning	Nature and expectations from summative assessment—assessment of learning
– Understand and use chemical terminology and rules for naming and communicating in chemistry – Plan and carry out experiments, estimate uncertainty and assess sources of error, present results and argue for the validity of results and conclusions – Use models to explain observations and chemical phenomena, and argue for the strengths and limitations of the models	The assessment during the course shall contribute to promoting learning and developing competence in chemistry 1. Students demonstrate and develop competence in the subject when they do experimental and exploratory activities, and when they use chemistry terms, terminology, models and calculations in explanations and reasoning. Furthermore, students demonstrate and develop competence in chemistry 1 when they reflect on and evaluate chemistry claims and information	The grade/score provided must be an expression of the overall competence the student has at the end of the training after chemistry 1. The teacher must plan and arrange for the student to demonstrate their competence in varied ways, including understanding, reflection and critical thinking, in different contexts. The teacher must grade chemistry 1 based on the competence the student has shown through practical, written and oral work in the subject. The point of view assessment must also be based on the competence the student has shown by using chemical terminology and forms of expression in work with academic subjects, contexts and issues

Note that the above table only shows a few selected competence goals. There is not enough space to include all the information provided in this curriculum. The purpose of including Table 6.1 is simply to show the connection between the list of chemistry knowledge and skills that could be assessed, which I provided earlier and how these are captured in a real chemistry curriculum. In case of interest in more details of this Norwegian chemistry curriculum, these can possibly be obtained by gaining access to (as of July 2024) the Norwegian directorate for education and training via their webpage—https://www.udir.no/lk20/kje01-02/kompetansemaal-og-vurdering/kv532

traditions which are a summary of two different schools of thought regarding teaching, learning and assessment practices. These are the *quantitative and qualitative* educational thinking traditions. These two educational thinking traditions are summarized in the next section, with a focus on how they respectively underly and thereby inform and influence the different assessment approaches existing in today's classrooms.

6.3.2 The Quantitative Versus Qualitative Educational Thinking Traditions

A distinction between the quantitative and qualitative educational thinking lies in how learning is conceptualized and hence how measurement of educational achievement is planned and executed. The quantitative tradition considers learning as "acquiring "specific discrete skills described as precise well-delimited behaviors" (Cole, 1990, p. 2), and the contents of learning as "discrete quanta of declarative or of procedural knowledge" (Biggs, 1995, p. 2). Such a conception of what counts as learning directly influences how curriculum is structured and organized as well as assessment practices. Biggs (1995) expresses that a curriculum that is underpinned by a quantitative educational thinking tradition is often "a list of discrete units: facts, skills, competencies, behavioral objectives, performance indicators, and the like, and assessment a matter of how many of these have been attained" (p. 2). The qualitative thinking tradition on the other hand moves away from the view that considers learning in terms of discrete packets of knowledge (quanta) and accounts for qualitative changes along the continuum of learning, where constructions the learner makes at any point in the learning process count as outcomes of the learning. Accordingly, the qualitative tradition is underpinned by constructivism or generally paricipationist/acquisitionist (Chinn & Iordanou, 2023) theories of learning as opposed to the quantitative tradition which has its roots in the positivist tradition (Biggs, 1995; Moss, 1992). Describing what counts as learning and what should direct assessment for the qualitative tradition, Biggs (1995) expresses that:

> ......understanding changes progressively as people learn, with qualitative changes taking place in the nature both of what is learned, and how it is structured. Understanding of a topic thus evolves cumulatively over the long haul, having "horizontal" interconnections with other topics and subjects, and "vertical" interconnections with previous and subsequent learnings in the same topic. ...... As the contents of learning are meanings, the curriculum question is to decide what meanings or levels of understanding are "reasonable" at the stage of learning in question. (p. 4)

6.3.2.1 Exploiting and Benefiting from Both Thinking Traditions

The interest in discussing the two educational thinking traditions as regards students' learning and assessment here in this didactics book, is not to focus on the shortcomings of these respective schools of thought. Rather, the focus or interest is about providing us with a wider perspective when it comes to designing, developing and or selecting plausible assessment strategies and tools for use in our classrooms. The aim is for us (teachers and educators) to relate these to our inherent thinking, teaching and learning experiences, and thereby make informed decisions about what kind of assessment practices we want to embrace to achieve the kind of envisioned educational achievement. That said, to summarize the distinction between the quantitative and qualitative educational thinking when it comes to assessment, Biggs (1995) expresses that, for the quantitative tradition, assessment "implies aggregating units of learning taken cross-sectionally with respect to time", while for the qualitative tradition, assessment "implies charting longitudinal growth over time, from relative ignorance to relative competence..." (p. 4). For us chemistry teachers and even teachers of other subjects, both conceptions of assessment are relevant and applicable in terms of establishing the level of educational achievement that we want to observe at the end of training. Undoubtably, our knowledge of theoretical conceptions and practices from both traditions guarantees wider possibilities for assessment strategies and tools, including the possibility of integrating ideas from both traditions when designing assessment tools.

6.4 Frameworks and Models of Assessment

There are several assessment frameworks and models presented in educational literature which could be relevant in chemistry classrooms. Most of these are largely rooted in the qualitative educational thinking tradition, though there are those that combine and use ideas from both traditions. In this section, I will select and discuss a few of these, namely, *(i) evidence centered design of assessments (*Mislevy & Haertel, 2006; Stowe & Cooper, 2019*); (ii) context-based assessment* (Bellocchi et al., 2016; Bortnik et al., 2021); *(iii) constructive alignment framework* (Biggs, 1996, 2014; Wang et al., 2013*); (iv) The KLT—theory of action framework—from ETS—(*Bennett, 2011; Wiliam & Thompson, 2017*), and (v) Bloom's taxonomy as an assessment guiding framework (*Furst, 1981*).* The selection here does not in any way imply that these are the most useful guiding frameworks for designing assessments. They are simply given as examples for you to consider and the decision whether to use and not to use the ideas and principles suggested in these respective frameworks remains entirely yours. This is the beauty of being a teacher, having the possibility to draw on a pool of available educational resources to develop and shape your own teaching.

6.4.1 Evidence Centered Assessment Design (ECD)

The Evidence-centered assessment design (ECD) is one such framework that provides for designing and implementing many kinds of assessments. Assessment according to Mislevy et al. (2003), is "an evidentiary argument: an argument from what we observe students say, do, or make in a few circumstances, to inferences about what they know, can do, or have accomplished more generally". ECD therefore "provides language, concepts, and knowledge representations for designing and delivering educational assessments, all organized around the evidentiary argument an assessment is meant to embody" (Mislevy & Haertel, 2006, p. 1). The ECD in its original details seems complex for a chemistry classroom teacher to embrace and employ in his or her day-to-day teaching. This is majorly due to complex language, terminologies and later alone the amount of details one needs to go through before grasping the operationalizable key ideas and principles of the

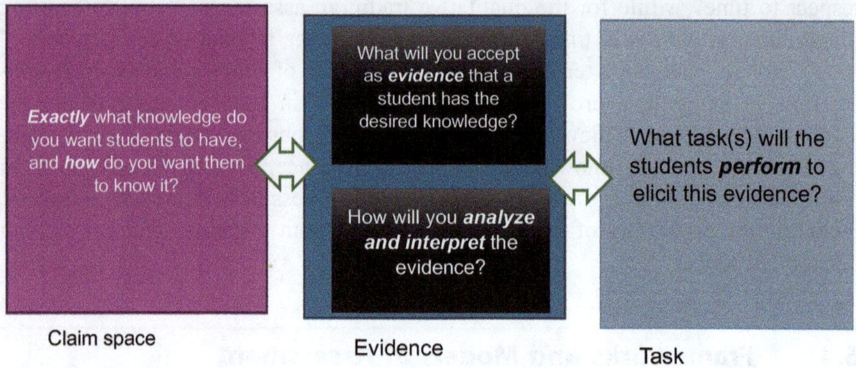

Fig. 6.1 The three step iterative ECD process as presented by Stowe and Cooper (2019)

model. Nevertheless, Stowe and Cooper (2019) managed to extract and summarize the central ideas of this framework and presented a more simpler version of the ECD model which seems applicable to chemistry teaching and assessment contexts. If we embrace and use Stowe and Cooper's conceptualization of the ECD framework, then applying the principles of ECD goes through three steps. First, precisely outlining what students should know and be able to do, second, stating the evidence that will meet expectations set out in step one and establishing how the stated evidence will be analyzed, and third, designing assessment tasks which have the potential to elicit the evidence stated in the second step. Figure 6.1 is a summary of these three steps of the ECD assessment model presented by Stowe and Cooper (2019).

Teachers can easily use the framework to design assessment tasks of different or varied cognitive requirements. Such tasks, for example, could be a combination of knowledge testing (recall questions) and questions of a conceptual nature such as the example shown in Fig. 6.2.

6.4.2 Context-Based Assessment (CBA)

A Context-based approach to assessment is one approach with support from several educational researchers (see for example; Bellocchi et al., 2016; Bortnik et al., 2021; Greeno et al., 2013; Ismail & Wiyarsi, 2020). It is often considered part of the well-known context-based learning approach, where contexts are used to promote students' understanding of science. Students' assessment tasks and activities that inform about their learning are tailored to problems and issues in real-life situations. That is, the assigned learning tasks are characteristic of activities performed in authentic scientific environments (Bortnik et al., 2021), and they are expected to specify and connect science concepts to situations familiar to students (Bellocchi et al., 2016). Different context-based assessment design frameworks can be found in educational literature; however, the underpinning principles are the same.

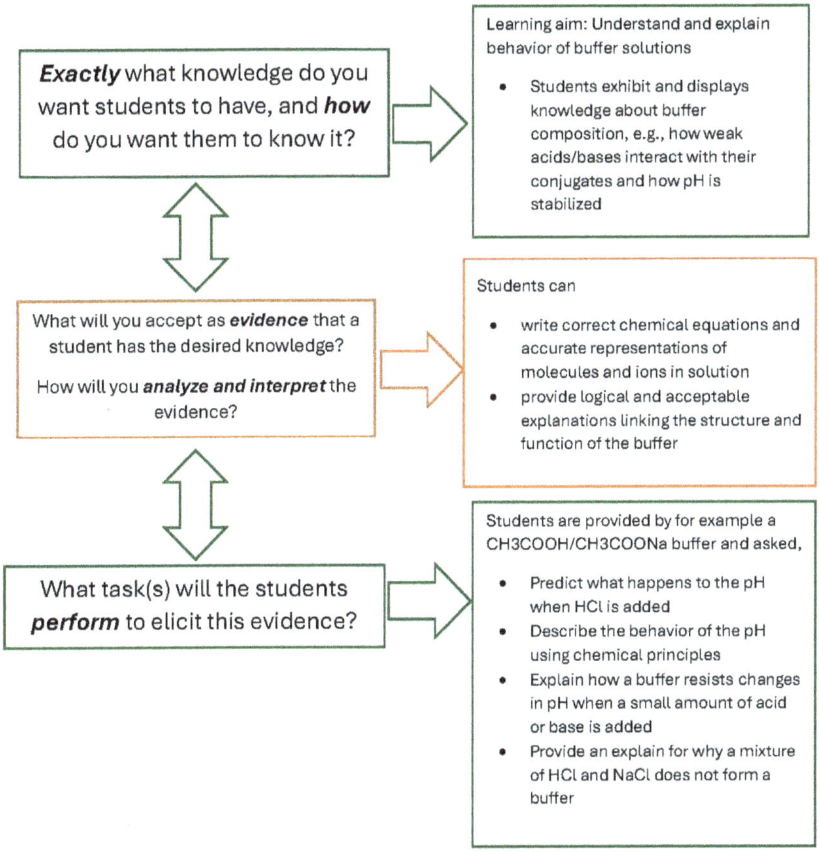

Fig. 6.2 Assessment design based on the ECD framework

That's, ensuring the clear connection between what is learned, the context being used in the learning and the tools, instruments or tasks being used to assess the intended learning. At least five features are characteristic of a CBA or situated assessment (Greeno et al., 2013), that is, (i) drawing on authentic contexts—mirroring real-world scenarios; (ii) using performance-based tasks that allow students to demonstrate skills and understanding through for example completing projects, engaging in case studies, simulations or real-life problems; (iii) considering the social and cultural aspects as key to students learning—for example taking into consideration learners' backgrounds and abilities when designing assessment tasks: (iv) the feedback to assessment activities is formative and dynamic—supporting students' learning over time as they work to achieve the target learning outcomes; (v) the CBA strategy combines and uses multiple data sources and tools—such as observations, reflections, portfolios, peer feedback, and self-assessment, as way to provide a holistic understanding of the students' progress and learning.

Any proposed framework for designing context-based assessment tasks would start by (i) considering and defining the goals and objectives of the intended assessment, which are linked to the overall goals and objectives of the study program. Note that there are several target learning goals in chemistry as earlier listed. Only a few of these can be assessed at a time depending on what teaching curricular aspects are being considered; (ii) choosing the subject area of focus (content of assessment) and their corresponding contexts, (iii) designing and sequencing the corresponding assessment tasks, and (iv) administering the assessment and analyzing, and using results in line with the goals of the assessment which could be formative or summative goals. For example, let us consider the well-known issue of water purification and safe drinking water to be a starting context that is indeed a real-world issue. There are several chemistry concepts or topics that can be linked to this water issue, including the topic of reactions in solutions—solubility and precipitation reactions., acid–base reactions and pH measurements or chemical contaminants and methods of water purification. Depending on how and what has been taught, the teacher can design context-based assessment tasks targeting students' understanding of aspects related to solubility and precipitation reactions, the application of acid–base titration and pH measurement, the evaluation of water quality and analysis of collected data as well as interpreting results. As mentioned above, the assessment task can consider and involve multiple data sources obtained using performance-based tasks.

6.4.3 The Constructive Alignment Teaching Framework (CA) and Assessment

A constructive alignment teaching framework (Biggs, 1996) is a teaching framework that emphasizes the connection between the intended or target learning outcomes, the assessment criterion that measures the attainment of these outcomes and the teaching and learning activities engaged in by both teacher and students, that which will result in the attainment of the intended learning outcomes. Accordingly, the teacher starts with determining and formulating the intended learning outcomes of the target lesson. Once these are well stated, then the second step is to develop the assessment criteria or regime based on the stated learning outcomes. The third step involves the teacher designing (formulating or organizing) activities that students will engage in to be able to meet the assessment criteria and hence the learning outcomes. The underlying understanding is that this three-step plan should be in place before any teaching commences (Biggs & Tang, 2015; Kandlbinder, 2014; Wang et al., 2013).

The CA teaching model is underpinned by a constructivist learning view that puts the learner at the center of learning. Leaning activities should be designed in a way that allows learners to construct their own knowledge and to work towards attainment of the target learning outcomes. The established assessment criteria or regime allows the teacher to provide feedback to students and provides for continuous improvement by students unto the attainment of the target learning outcomes. The mode of assessment within the CA model is open and apart from being framed within a constructivist teaching philosophy, the teacher is free to decide what kind of assessment tools s/he will employ, which will inform on the attainment of the target learning outcomes. In the planning of the entire framework before teaching, there are three guiding questions, (i) what are the target (plausible) learning outcomes for this module that I plan to teach? (ii) how will I establish that my students have attained or achieved the intended/target learning outcomes? (iii) What kind of activities are suitable, that I (the teacher) and students could engage in, that which meet my assessment criteria and hence support students in achieving the intended learning outcomes?

6.4.4 The KLT—Theory of Action and Framework for Formative Assessment

Bennett (2011), in a critical review of formative assessment, argues that a meaningful definition of formative assessment requires a theory of action and a concrete instantiation (p. 8). Accordingly, the theory of action characterizes and qualifies the components of formative assessment, and how these components work together to create a set of desired outcomes. The instantiation on the other hand "illustrates what formative assessment built to theory looks like and how it might work in a real setting." (p. 8). Bennett offers an example of a theory of action to illustrate his conceptions with elements that are more understandable to a classroom teacher and hence transferrable to teaching situations. It is this theory of action that I highlight in this part, that I find to provide a good basis to think about and plan for formative assessment in chemistry classrooms. Bennett extracts what he calls a rudimentary theory of action from the development work of the "Keeping Learning on Track KLT" program (Thum et al., 2015; Wiliam & Thompson, 2017), which he describes to comprise *one big idea* and *five strategies key strategies*. The big idea entails teachers and students using evidence about learning to adjust the teaching to meet students' needs. The five key strategies on the other hand include sharing learning expectations, questioning, feedback, self-assessment, and peer assessment. Accordingly, the strategies are used to establish where students are in the learning process, where they are going and how they will get where they are going (Bennett, 2011; Wiliam & Thompson, 2017). Wiliam and Thompson (2017), present the five key strategies in a tabular form as shown below in Table 6.2, in a way that enables visualizing the relationship between these key strategies and a formative assessment roadmap.

Table 6.2 The five key strategies of formative assessment as summarized by Wiliam and Thompson (2017)

	Where the learner is going	Where the learner is right now	How to get there
Teacher	Clarifying learning intentions and criteria for success	Engineering effective classroom discussions, questions, and learning tasks that elicit evidence of learning	Providing feedback that moves learners forward
Peer	Understanding learning intentions and criteria for success	Activating students as instructional resources for one another	
Learner	Understanding learning intentions and criteria for success	Activating students as the owners of their own learning	

6.5 Bloom's Taxonomy as an Assessment Guiding Framework

The taxonomy of educational objectives is a framework for classifying statements about what is expected to be learned or intended to be learned by students as a result of instruction (Anderson et al., 2001). The original framework by Benjamin S. Bloom and colleagues published in 1956 comprised three domains, the cognitive, the affective and psychomotor domain. The idea behind its development was not for the framework to be used only for measurement purposes but also as a framework that serves as a common language for communicating about learning goals across subjects, a basis for curricula development and review, a basis and a means for connecting educational objectives, teaching activities and assessment, and as a guiding framework for quality assurance of educational programs (Bloom et al., 1956). The cognitive domain, however, was rapidly considered suitable for classroom learning measurement purposes, and hence a framework guiding the development of test items targeting different cognitive levels. Bloom's (1956) original version of cognitive domain comprised six levels of thinking, knowledge, comprehension, application, analysis, synthesis, and evaluation. These levels were ordered hierarchically from simple to complex and from concrete to abstract in terms of cognitive demand, where mastering one level is considered a prerequisite to mastering the next level. The cognitive domain categories were later revised by Anderson and his colleagues into six categories, remember, understand, apply, analyze, evaluate, and create (Anderson et al., 2001).

Over the years, Bloom's cognitive domain categories have been employed to guide educators at all levels and across disciplines in designing assessments items for different purposes. There is no doubt about the fact that the framework is still useful even in science classrooms today. It is a truism that cognitive occurrences within a learner's mind cannot be directly observed, and this was one of the main criticisms of Bloom's taxonomy, in addition to a criticism against organizing cognitive thinking hierarchically (see for example; Gall, 1970; Kayima, 2018). On one hand, the categories appear simple for the teacher to consider the kind of question items that suit a desired assessment level. On other hand, however, it can be difficult to fully establish whether the understanding displayed by the learners at any level along Bloom's taxonomy indeed translates into learning at that level. This is because it is also possible to memorize and produce knowledge at advanced levels without understanding the concepts themselves or the principles underlying the concepts of interest. This was also Bloom's (1956) concern when he acknowledged that it is not possible to know whether a student who has been asked a particular question that requires a high-level cognitive process, indeed uses a high -level cognitive process to answer the question. Nevertheless, the framework can work as a starting point for chemistry teachers, and in particular the novice ones who are still learning to develop assessment items to serve different cognitive demands.

As an example, let us use some information from Table 6.1, that is, one of the three target competence goals provided in the first column and then position these with Bloom's cognitive levels. The teacher can for example state that "my students will display their competence development when they are able to answer question items along Bloom's taxonomy". The teacher then can start to develop question items corresponding to each one of the six cognitive levels. Remember, as I have already mentioned above, there is no directly observable connection between the student cognitive processing (engagement) and the question given to the student to answer. We can only presuppose this ongoing cognitive activity from the outside, for example by considering the amount of time the task requires, how many steps to go through to solve the question and what resources the student needs, whether answering the question demands making connections, reflections and thinking, etc., and by placing ourselves in the student's shoes and imagine ourselves solving the same tasks and use this to establish the kind of demands the question will pose.

6.6 Possible Assessment Forms, Methods, Strategies or Instruments

There is a myriad of assessments strategies described in educational literature for chemistry teachers to choose from when assessing their students in chemistry. Any form or strategy of assessment or instrument can be used to assess students formatively (assessment for learning) or summatively (assessment of learning) if the teacher has a well-defined plan and structure for how to use the instrument to achieve the set-out assessment goals. In other words, the success of any assessment relies on the teacher having a well-established assessment plan with well-defined goals for this assessment, where the following three questions are answered,

i. What is the purpose of the assessment?—Why assess?
ii. What are the contents of the assessment?—What exactly will be assessed?

iii. What is the most suitable instrument or strategy for achieving the intended assessment purpose?—What kind of methods (tools) are most suitable to for intended assessment?

At the center of any assessment are the teacher's assessment competences and preparedness. This does not imply that teachers must categorically exhibit special assessment skills to be able to successfully assess their students. On the contrary, it implies the freedom teachers have in deciding what suitable assessment strategies or instruments with which they will support their students towards attaining target learning outcomes (assessment for learning), and with which they will seek to establish the level of attainment of the target learning outcomes (assessment of learning). Moreover, teachers themselves will find out that some strategies or instruments are more suitable for formative assessment while others serve better as summative assessment tools, strategies or instruments. In Table 6.3, I have collected a list of some of the assessment methods, strategies, tools or instruments potentially useful or relevant in chemistry classrooms. As mentioned, determining what assessment method, tool, strategy or instrument serves better and where, is the teacher's responsibility. The teacher must examine the suitability of the selected or designed assessment strategy for the intended assessment. The list presented in Table 6.3 is for guiding purposes only.

6.6.1 Specific Case Examples of Assessment Tasks in Chemistry

In this section, random examples of assessment tasks are provided. The examples are meant to show that it is possible to formulate and administer different tasks and that the teacher is free to explore and use any plausible approach or instrument as shown in Table 6.3, if s/he has well-defined intentions (purpose or aim) for the assessment. Note that planning for assessment should normally consider other factors such as the curriculum goals, timeframe available, expectations from stakeholders, etc. In other words, assessment, whether formative or summative, is not an isolated classroom activity but one part of the teaching and learning complex and must be carried out within the framework of this complex.

6.6.1.1 Case Example 1: Goal of Assessment: Assessment for Learning

Assessment strategy:	True or false statements
Goal of assessment:	Assess knowledge of the underlying basic chemical principles; the ability to pay attention to small details

(continued)

(continued)

Assessment strategy:	True or false statements
Task and instructions	The information provided is a mix of both correct and incorrect statements, expressions, symbols and formulae a. Carefully read through the entire text, identify and correct all statements, expressions, symbols or formulae that you deem incorrect as per your current understanding, rewrite and present the information in the way that you deem correct b. Write a short account of every identified incorrect statement, symbol or formulae to justify why you deem it incorrect
Task contents[3]	Acids and bases are two important classes of chemicals that have completely opposite properties. Acids are substances that accept protons, while bases are substances that donate protons. This is known as the Lewis theory of acids and bases. One of the most common acids is hydrochloric acid (HCl), which is found in car batteries. It is a strong acid, meaning it completely dissociates in water to produce H^+ ions and Cl^- ions. Another example of a strong acid is acetic acid, which is found in vinegar. On the other hand, ammonia (NH_3) is a classic example of a strong base. It dissociates completely in water to form NH_4^+ and OH^- ions. Sodium hydroxide (NaOH) is another strong base, which is commonly used in household cleaning products Bases can be identified by their sour taste and their ability to turn blue litmus paper red. Conversely, acids are known for their slippery feel and their ability to turn red litmus paper blue. The strength of an acid or base is measured by its pH level. The pH scale ranges from 0 to 14, with 7 being neutral. A pH value below 7 indicates a base, while a pH value above 7 indicates an acid. For example, lemon juice has a pH of about 2, making it a weak base, whereas bleach has a pH of about 12, making it a weak acid In an acid–base reaction, the acid and base neutralize each other to form water and salt. This is called a neutralization reaction. For instance, when hydrochloric acid reacts with sodium hydroxide, the products are water and sodium chloride, which is table sugar Understanding acids and bases is crucial in various fields such as chemistry, biology, and environmental science. They play a key role in metabolic processes, industrial applications, and maintaining the ecological balance

The task in case example 1 presents a lot of opportunities for teachers to engage their students either on an individual basis or as a whole class. The text is full of incorrect statements and claims and depending on how each student performs, the teacher may decide the kind of feedback suitable for each student. Moreover, the provided case example is equally suitable for summative assessment if the teacher has well-defined intentions for the assessment. For this assessment to be successful, it must happen within a specified timeframe and students should not have access to any other information that might jeopardize the essence of the activity. The underlying assumption is that students have already studied the concepts presented in the assessment task. Alternatively, the assessment can be expanded by providing students with reading materials covering the topic for self-study, at least

[3] The task content is ChatGPT generated.

Table 6.3 List of possible assessment forms, methods, tools, strategies or instruments

Methods, tools, strategies or instruments	Short description	AFL[4]	AOF[5]
Classroom observations	Teachers can observe not only how students behave and participate in classrooms but can keep track of what students are doing as basis for productive feedback and for mapping transitions in students learning. Students displayed skills, knowledge and transformed attitudes are case examples of where classroom observations can be useful	x	
Informal/formal conversations * Teacher and student *Teacher and parent	Some schools have a formal structure where the teacher meets each student's parents to discuss both the status and learning opportunities going forward. Students can also meet the teacher to get individual feedback and support on how to progress	x	
Use of oral questioning and traditional pencil-and -paper tasks (quizzes)	The use of oral questioning helps facilitate class discourse between teacher and student. Teacher diagnoses students' learning needs through questioning and provides necessary support With Quizzes, there is an opportunity to actively enjoy the lesson while reflecting on one's strengths and weaknesses on the part of the students. For the teacher, it allows for identification of progress and where support or attention should be put among other things	x	x
Video assessment	Use of visual recordings to provide assessment feedback to students. Students can also use video recordings to give feedback on teaching	x	
Surveys and polls	Can be used the teacher to gain insight into students' learning, mastered skills or attitudes relating to learning	x	
Exit laps/cards	Students are asked to write down at least three main points or interesting aspects at the end of the lesson and at least one question about what they still wonder about or that which they are yet to understand	x	
True or false statements	Statements are provided and students are free to use any available means to verify them as one way to activate their thinking	x	

(continued)

[4] AFL—Assessment for Learning—the column is checked to indicate the suitability of the instrument/method/tool/strategy in question.

[5] AOF—Assessment of Learning—the column is checked to indicate the suitability of the instrument/method/tool/strategy in questions.

Table 6.3 (continued)

Methods, tools, strategies or instruments	Short description	AFL	AOF
Self and peer assessment *E.g., Use of Consolidation activities for self-assessment	Peer assessment gives the opportunity to students to look at what their fellow students have done, reflect on it, and provide feedback based on a pre-established criterion. In so doing, the student takes the reflection back to her/his own work and extract understandings that s/he might not have before the assessment Both self and peer assessment should provide the students the opportunity to examine their work, reflecting on their product and the reasons for their current thinking and thereby recognize strength of their work and areas for improvement	x	
Learning logs	Documentation of what the student is doing, what the student has accomplished and what is pending including ideas and questions	x	
Home assignments	A take home or work from home assignment. Students are given tasks of varying complexity to work on or solve while at home, submit within a specified time and get feedback from the teacher	x	
Home exams and open book examinations	With open book and home examinations—students are allowed to use any available resources—textbooks, notes, etc., to answer		x
Mobile-based		x	
Use of digital technologies and tools *Game-based *Kahoot *Online problem-based learning *Computer assisted scoring	There are number of developed digital assessment resources and tools. Teachers are expected to orient themselves, constantly update their knowledge and skills to be able exploit and benefit from using these resources	x	
Written tests and exams	Used to assess students' knowledge and the attained competences in chemistry. Teachers can formulate questions of varied cognitive demands, ranging from recall questions to higher-order conceptual questions. Bloom's cognitive categories are often used as a guide when constructing test items		x

(continued)

Table 6.3 (continued)

Methods, tools, strategies or instruments	Short description	AFL	AOF
Reports (from experiments, fieldwork and excursions)	Lab report—students conduct chemistry experiment either individually or in groups and thereafter write a report. The contents expected in the report are normally provided to the students. Depending on the nature of the experiment, different aspects of knowledge, skills and attitudes can be assessed. E.g., the ability to observe, record and interpret results Field work/excursions report—like the lab report, it's the final product submitted after an excursion or field work. The teacher provides the scope and expectations of the contents of the report	x	x
Oral presentations	Students can display their understanding through oral presentations. Can be used for assessment of students learning and for assessment for learning	x	x
Project work			x
Work-books	This is the day-to-day student's notebook with tasks, assignments or exercises of different cognitive levels aligning with the planned teaching that goes on in the classroom. It is normally provided as a supplement to the textbook to complement and reinforce students' learning of topics in the textbook	x	
Multiple choice questions (MCQs)	Are fixed response questions comprising a question stem and a choice of answers—There are several types of MCQs ranging from MCQs with one attractor (correct response) and several distractors (incorrect choices), MCQs with more than one attractor, true or false, ordered multiple choice questions, complex multiple-choice questions, etc	x	x
Group work			x
Closed response tasks	Questions with pre-defined or known answers		x
Open response questions *Short answer questions	Students' responses are not constrained but the teacher could determine the scope of the response, that is, ranging from having short response answers to more extended responses	x	x
Performance assessment (PAs) *Practical work and extended projects *Oral examination *Web-based video responses	PAs allow students to demonstrate and apply their acquired knowledge, thinking processes and skills as they complete performance tasks. The tasks target students' application of knowledge and skills related to a set of defined learning outcomes. Different performance tasks can be designed such as practical activities, extended projects, oral examinations, tasks that require students to interpret media reports, review or examine science reports, etc	x	x

(continued)

Table 6.3 (continued)

Methods, tools, strategies or instruments	Short description	AFL	AOF
Criterion referenced (CRT) Vs Norm referenced (NRT) testing	CRT—assessment tasks are based on a pre-determined standard or criterion often made available to students at the time of the test. NRT—assessment test compares a student's score to a representative group and reports the student's performance with respect to the representative group		x
Comparative judgment	Two or more performances are compared to decide which one of them is better		x
Interviews	Interviews can be used to allow for a one-to-one formal or informal exchange between teacher and students. Moreover, to some students, it gives the student the chance to express him/herself (regarding their understanding) without being afraid of being judged by fellow students	x	
Science journal entries	Guided by the teacher, students can keep a record of their science learning activities as they progress, including a record of observations, ideas and thoughts about science in nature, in books and in classrooms, etc. The journal can be a basis for self-assessment and for tracking transitions in knowledge development	x	
Rubrics/checklists	Teachers can design competence, knowledge or skills rubric or checklist. It could also be a learning outcomes rubric or checklist. With this, s/he can proceed to design assessment activities seeking to measure the extent to which students have attained or mastered a given target competence or skill. A constructive alignment framework as described earlier could be suitable as a planning framework here	x	x
Visual displays—*posters; *concept maps; *models, *Gallery walks *PowerPoint presentations	Visual displays allow for processing available or explored information and producing a framework that represents derived understandings *Concept maps—A diagram that students use to visually represent a concept. It helps students organize and visualize their knowledge, they make and show connections between ideas *Gallery walks—students prepare posters or concept maps about certain concepts and display these. Fellow students move from one poster to another as the respective poster owners describe and defend their displayed understandings	x	x
Drawing in science	Students represent their understandings through drawings. Teachers use the opportunity to examine students' meanings, discover misconceptions and support further understanding	x	

a week before the actual testing. The latter is mainly if the goal of the assessment is a summative one. Moreover, when you look through the list provided in Table 6.3, you can find that there are several ways that a teacher can use to provide feedback following students' completion of the task. For example, it is possible to observe students as they complete the task and base the feedback on the process students go through in the attempt to identify the errors and incorrect statements as demanded in the task. The teacher can choose to have a one-to-one talk with the student to understand the student's line of thinking and provide support as needed, or the teacher can also decide to provide feedback through video recording, etc.

6.6.1.2 Case Example 2: Goal of Assessment: Assessment of Learning

Assessment strategy:	Closed response task
Goal of assessment:	Understanding and application of the basic principles
Task contents	Element X has two stable isotopes X_1 and X_2. Let us assume that r_1 and r_2 are the respective mass numbers of the two Isotopes of Element X, whereby $r_2 - r_1 = 2$. If q is the total number of protons that Element X has, i. How would the expression for the total number of neutrons look like for both isotopes X_1 and X_2? Could you state the reason why the two expressions are as you have stated them? ii. If r_1 is five times the square root of 49, what is the value of r_2 considering the relationship between r_1 and r_2 given above? iii. Suppose that q equals 17, use all the information you have gathered so far from your responses to questions i and ii above, to solve the expressions and identify Element X iv. Construct or write a short but coherent story (between 200 and 1000 words) about Element X using at least nine of the following key words, nature, isotopes, abundance, valence, alkali metals, salts, halogen, bonding, ionic, reactive. You are also free to include any other terms or concepts to make your story coherent

The task in case example 2 is also suitable for both formative and summative assessment purposes. The task is very simple, requiring a simple recall of the relationship between mass number, atomic number, and the total number of neutrons of a given atom. The task uses letters and simple mathematical expressions to challenge the student's cognitive processing and his/her ability to hold on to the basic principles and not being distracted by the mathematical expressions and the used substitute letters. Such tasks can be constructed to challenge students in such a way that the task demands forward and backward thinking in the process of finding an answer to the task. We can consider this form of cognitive engagement as opting for the alternative, bumpy and long route instead of the short and direct route to the destination. The teacher can create a task with various obstacles that the student must deal with first before solving the actual task. Think about this in the form of solving a game puzzle, where the gamer must connect or put together different bricks prior to getting the puzzle correct. The process towards solving the task itself will inform on the student's learning, where s/he is in the learning process and where the student might need support to succeed in going forward.

6.6.1.3 Case Example 3: Using Video Feedback to Assess Learners Formatively (Assessment for Learning)

During my several years as a science teacher educator, I have supervised several groups of science teacher students who have explored the topic of assessment during their internship (school practice), and as part of their practice-based learning. The way teacher training is organized is different for different institutions across different countries. At one of the teacher-training institutions I have worked, teacher students have had the opportunity to design action-based mini-projects and to carry out these mini-projects during their school practice. The aim is to allow these students to consider a real teaching problem and to find a plausible solution to this problem and in the process, learn to use their knowledge from teacher training in real teaching contexts and thereby shape and improve their teaching competences.

Of the several groups of teacher students I have supervised during their school practice, one group planned and carried out a qualitative action-based mini-project, where the goal was to explore and establish whether using video-feedback would be a better approach to formative assessment that promotes students' learning. The mini project was designed such that the group first carried out classroom teaching where they provided feedback in the known traditional way, namely asking oral questions to check whether students have understood or not as the lesson progressed, giving working tasks at the end of the lesson, correcting and giving written comments to these tasks. After this first round of teaching and traditional formative assessment, the group then started a new round of teaching with the same students but using an alternative assessment approach, namely video-based assessment. Accordingly, the group used what is termed as "Screencast feedback" (Mahoney et al., 2019) which involves recording the teacher's computer screen showing a mouse that moves around as the teacher is scrolling up or down the captured students' work and wherever needed typing or pointing to where attention is needed, simultaneously as he/she talks (audio narration). With Screencast feedback, the teacher's image or photo is not the focus and may not be included. The focus is on students' work, normally presented in the form of a moving image simultaneously with the teacher's voice explaining whatever the center of focus is at any given instance.

In both rounds, the group engaged their students in a semi-structured interview after each round of teaching and assessment, to obtain information about how the alternative assessment approaches have respectively contributed to students' progressive learning going forward. Combining the results of the interviews, insights and experiences gained throughout the whole action process, the group made several reflections and conclusions regarding the use of video feedback as one form of formative assessment to promote students' learning. Among other things, the group found that their students considered video feedback to be thorough, concrete, rich in details, readily available, and easy to understand. Moreover, the group noted that their students were more likely to revisit and use the video feedback as many times as could be needed. Moreover, the visual component of this kind of feedback seemed to attract students' attention. On the other hand, the group noted that, like the written feedback, video-based feedback did not or does not provide the opportunity of a dialogue between teacher and students, unless the teacher is willing to organize an extra dialogue session. Seen in the eyes of a teacher, this is typically a question of work overload and hence less attractive as a means of formative assessment.

From the short story about the action-based mini-project my students carried out, using video back is possible in chemistry classrooms and can be one of the ways to promote students' learning. Moreover, video feedback could be an option for those students who are "considered" weak in the subject, while at the same time, are shy or afraid to reach out for help even when face-to-face sessions are offered. In some cases, the teacher can use the video recording to reach out to the student in question with a tone that results in a change of attitude towards the subject. Note that with the Screencast feedback, the teacher's image is often not shown as the focus is on the student's work. A teacher who wishes to convey a form of video assessment that includes both verbal and nonverbal expressions—facial and body language, can use a combination of the Screencast feedback and what is termed as "talking head video feedback" (Lamey, 2015), where the teacher records him/herself talking about the student's work. A combination of these two implies the student can both see his/her own work as it is being commented on and the teacher who is giving the feedback.

References

Anderson, L. W., Krathwohl, D. R., Airasian, P. W., Cruikshank, K. A., Mayer, R. E., Pintrich, P. R., Raths, J., & Wittrock, M. C. (2001). *A taxonomy for learning, teaching, and assessing: A revision of Bloom's taxonomy of educational objectives*. Longman.

Bellocchi, A., King, D. T., & Ritchie, S. M. (2016). Context-based assessment: Creating opportunities for resonance between classroom fields and societal fields. *International Journal of Science Education, 38*(8), 1304–1342. https://doi.org/10.1080/09500693.2016.1189107

Bennett, R. E. (2011). Formative assessment: A critical review. *Assessment in Education: Principles, Policy & Practice, 18*(1), 5–25. https://doi.org/10.1080/0969594X.2010.513678

Biggs, J. (1995). Assessing for learning: Some dimensions underlying new approaches to educational assessment. *Alberta Journal of Educational Research, 41*(1), 1–17.

Biggs, J. (1996). Enhancing teaching through constructive alignment. *Higher Education, 32*(3), 347–364. https://doi.org/10.1007/BF00138871

Biggs, J., & Tang, C. (2015). Constructive alignment: an outcomes-based approach to teaching anatomy. In L. K. Chan & W. Pawlina (Eds.), *Teaching anatomy: A practical guide* (pp. 31–38). Springer International Publishing. https://doi.org/10.1007/978-3-319-08930-0_4

Biggs, J. B. (2014). Constructive alignment in university teaching. *HERDSA Review of Higher Education, 1*(5), 5–22.

Bloom, B. S., Engelhart, M. D., Furst, E. J., Hill, W. H., & Krathwohl, D. R. (1956). *Taxonomy of educational objectives: The classification of education goals: Handbook 1: Cognitive domain* (Vol. 1). David McKay.

Bortnik, B., Stozhko, N., & Pervukhina, I. (2021). *Context-based testing as assessment tool in chemistry learning on University level* [450]. Basel, Switzerland.

Chinn, C. A., & Iordanou, K. (2023). Theories of learning. In N. G. Lederman, D. L. Zeidler, & J. S. Lederma (Eds.), *Handbook of research on science education* (pp. 89–120). Routledge Taylor & Francis Group.

Cole, N. S. (1990). Conceptions of educational achievement. *Educational Researcher, 19*(3), 2–7.

Furst, E. J. (1981). Bloom's taxonomy of educational objectives for the cognitive domain: Philosophical and educational issues. *Review of Educational Research, 51*(4), 441–453. https://doi.org/10.3102/00346543051004441

Gall, M. D. (1970). The use of questions in teaching. *Review of Educational Research, 40*(5), 707–721. https://doi.org/10.2307/1169463

Gibney, D. (2018). Towards an ideal chemistry curriculum. *School Science Review, 100*(370), 30–35.

Greeno, J. G., Collins, A. M., & Resnick, L. B. (2013). Cognition and learning. In D. C. Berliner & R. C. Calfee (Eds.), *Handbook of educational psychology*. Routledge. https://books.google.no/books?id=yw8FmsvgzuQC

Ismail, C. D., & Wiyarsi, A. (2020). Developing a context-based assessment: Task characteristic facet. *Journal of Physics: Conference Series, 1440*(1), Article 012015. https://doi.org/10.1088/1742-6596/1440/1/012015

Kandlbinder, P. (2014). Constructive alignment in university teaching. *HERDSA News, 36*(3), 5–6.

Kayima, F. (2018). *Exploring chemistry teachers' perspectives on questioning and providing a new way of analyzing teacher questions in science classrooms.*

Lamey, A. (2015). Video feedback in philosophy. *Metaphilosophy, 46*(4–5), 691–702. https://doi.org/10.1111/meta.12155

Mahoney, P., Macfarlane, S., & Ajjawi, R. (2019). A qualitative synthesis of video feedback in higher education. *Teaching in Higher Education, 24*(2), 157–179. https://doi.org/10.1080/13562517.2018.1471457

Mislevy, R. J., & Haertel, G. D. (2006). Implications of evidence-centered design for educational testing. *Educational Measurement: Issues and Practice, 25*(4), 6–20. https://doi.org/10.1111/j.1745-3992.2006.00075.x

Mislevy, R. J., Steinberg, L. S., & Almond, R. G. (2003). Focus article: On the structure of educational assessments. *Measurement: Interdisciplinary Research and Perspectives, 1*(1), 3–62. https://doi.org/10.1207/S15366359MEA0101_02

Moss, P. A. (1992). Shifting conceptions of validity in educational measurement: Implications for performance assessment. *Review of Educational Research, 62*(3), 229–258.

Stowe, R. L., & Cooper, M. M. (2019). Assessment in Chemistry Education. *Israel Journal of Chemistry, 59*(6–7), 598–607. https://doi.org/10.1002/ijch.201900024

Thum, Y., Tarasawa, B., Hegedus, A., Yun, X., & Bowe, B. (2015). Keeping learning on track: A case-study of formative assessment practice and its impact on learning in Meridian School district. *Northwest Evaluation Association.*

Wang, X., Su, Y., Cheung, S., Wong, E., & Kwong, T. (2013). An exploration of Biggs' constructive alignment in course design and its impact on students' learning approaches. *Assessment & Evaluation in Higher Education, 38*(4), 477–491.

Wiliam, D., & Thompson, M. (2017). Integrating assessment with learning: What will it take to make it work? In C. A. Dwyer (Ed.), *The future of assessment: Shaping teaching and learning* (pp. 53–82). Routledge.

7 Practical Work in Chemistry

Abstract

The way the term practical work is conceptualized and used in science education has changed over time from a focus exclusively on laboratory (in class or out-of-class) experiments—laboratory techniques and procedures to a more broader and encompassing conception, which includes any teaching and learning activity where the involved students, individually or as groups, observe or manipulate objects and materials with the aim of developing some kind of knowledge, skill or understanding. The focus of this chapter is on practical work activities often carried out in the form of experiments and the knowledge, skills, as well as preparations teachers need to make to be able to exploit the potential learning benefits of practical work in chemistry.

7.1 Practical Work in Chemistry: What Is It About?

The way the term practical work is conceptualized and used in science education has changed over time from a focus exclusively on laboratory (in class or out-of-class) experiments—laboratory techniques and procedures to a more broader and encompassing conception, which includes any teaching and learning activity where the involved students, individually or as groups, observe or manipulate objects and materials with the aim of developing some kind of knowledge, skill or understanding (Abrahams & Reiss, 2012; Gott & Duggan, 1996; Hodson, 1998; Millar & Abrahams, 2009). Moreover, while some educators seem to disagree (e.g.; Abrahams & Reiss, 2012), some others also consider activities such as computer simulations, modelling, and role plays, etc., a constituent of the "practical work" umbrella (e.g.; El Kharki et al., 2020; Pereira et al., 2020). Thus, if asked today (July 2025) to describe what constitutes the term "practical work", this latter conceptualization would be more acceptable as a more open-ended and encompassing term that gives the teacher or researcher the opportunity to think

F. Kayima, *Preparing for Chemistry Teaching*, Springer Texts in Education,
https://doi.org/10.1007/978-3-031-97122-8_7

about science practical activities far beyond the school laboratory walls. With this broader definition, the science educational research community describes practical work in science as varying along a continuum from closed-ended teacher-driven to the open-ended learner-driven practical activities (e.g.; Akuma & Callaghan, 2019b; Kidman, 2012; Ottander & Grelsson, 2006). However, in this chapter, focus is put on those practical work activities often carried out in the form of experiments, where either the teacher or students practically explore either the behavior, transformation, formation, or existence of varied forms of matter (chemicals), as a way to better understand its composition and properties, how and why matter undergoes change.

7.2 Why Do We Have or Talk About Practical Work in Chemistry?

Quite often when chemistry classroom teachers are asked to respond to the question of why science or chemistry experiments are an important component of the science or chemistry curriculum for different educational contexts, or why we have practical expriments in science or chemistry classes, the obvious response normally given is "because chemistry or science is a practical subject". It appears the teachers are reluctant to delve themselves deeper into the underlying historical grounds against which the existing interpretive frame for how we teach science is built. Indeed to date (July, 2025), science education research reports that chemistry teachers embrace the goals for practical work in chemistry as advanced in science education literature without themeselves questioning the role of practical work in chemistry. Studies investigating teachers' practices associated with the practical work component have reported that many teachers carryout practical work activities in chemistry simply because this is a normalized tradition, and because they understand chemistry to be a practical subject (see; Abrahams & Reiss, 2012; Millar et al., 2002; Toplis, 2012; Wellington, 2002). Furtherstill, studies looking at what reasons inform teachers' choices of practical activities indicate that teachers' decisions are mostly informed by their understanding of the nature of chemistry (the ontological component of chemistry) as well as psychological considerations - relating to their understanding of the teaching learning process (Lewthwaite, 2014). We must admit as science or chemistry teachers that our justification for the place and role of practical work in science should and must go deeper than a display of a precedural teaching practice that we have embraced without questioning. We need to understand and exactly answer the question of how we got to practice chemistry teaching the way we do it today, and what lies behind our conceptualization that chemistry is a practical subject that must be taught in ways that include students engaging in practical activities.

Science education as a broad discipline has evolved over time and with it a globally accepted framework for science teaching. While on one hand, the nature of science teaching has been mostly influenced by theoretical conceptions from (educational) psychologists regarding how pupils learn (e.g.; Dewey, 1929; John-Steiner & Mahn, 1996; Piaget, 1980; Vygotsky, 1962), on the other, the nature of science itself, has played a key role in shaping a framework for science teaching and learning (see for example; Gott & Duggan, 1996; Gunstone, 1991; Wellington & Ireson, 2013). Proponents of these two conceptual understadings about the process of science teaching and learning, consider practical work to be a central component with a distinct role in the science teaching and learning process. One way to understand and describe the place and goals for practical work in science in general and chemistry in particular, is thus to understand these two underlying theorectical conceptions that inform the science teaching and learning process.

Generally, about four main goals for practical work are mentioned in science education research literature, (i) understanding scientific concepts, (ii) arouse students' interest and motivation, (iii) train and acquire practical skills and problem solving abilities, and (iv) attain basic knowledge about the nature of science (scientific knowledge, method and processes) (see for example; Abrahams, 2011; Hodson, 1990; Hofstein & Lunetta, 2004; Lazarowitz & Tamir, 1994; Wellington & Ireson, 2013). The above four target aims are clearly meant to support students' learning of science, and are underpinned by how the nature of science is conceptualized, as well as the theoretical conceptions (learning theories) about how learning takes place. As an example, let us consider Alex Johnstone's (1982, 1991) earlier conception of the chemistry triplet (the macro, submicro and symoblic levels) and how this understanding shaped the nature of chemistry teaching the years that followed. Accordingly, chemistry comprises two worlds, the molecular and the macro worlds. Access to the molecular world is limited and scientists have worked very hard through several years and at different timelines to construct a form of medium with which we can communicate and exchange information and understandings of the interactions between different forms of matter. This third aspect or dimension of chemistry pertains its langauge, comprising the symbols, the numbers, equations, and other mathematical models and representations (e.g.; Cooper & Stowe, 2018; Johnstone, 1982, 1991). One of the central arguments emerging from Johnstone's conceptions was that, learning chemistry was difficult for students as it demands a multi-level thought engagement. That is, "students run into into trouble when teachers conduct teaching at the middle of the triangle, consdiering all three levels similitenously or when a teacher primilary concentrates at the symbolic level" (Johnstone, 1991). One way to deal with this difficulty is considering first hand experiences such as experimetal demostrations involving observations and manuplation of equipment at the macro level. The integration of practical aspects of chemistry was considered to play a substantial role in the students' understading of what happens at the submicro level, and to make it easier for students to make connections between all three levels. As such, the nature of chemical knowledge (the ontological chemical knowledge as described in chapter one) itself and how the process of human learning has been conceptualized, are central elements in justifying the place and goals for practical work in science teaching in general and chemistry in particular.

Moreover, the science education community also strongly believe that the goals for practical work in science extend beyond the focus on learning science knowledge, skills, and processes involved in science knowledge construction (e.g.,; the scientific method), and include a society dimension (see e.g.; Mahaffy, 2004, 2006; Sjöström, 2011). The focus here is put on the plausible connections between science technical skills acquired in classrooms and the techical skills applied at job placements—industries, medical laboratories, etc., including sustainability skills and practices as described in Chapter 10. Some important questions often discussed in some science education research reports, and which are relevant to us as teachers include, Do students benefit from understanding the nature of science, or when they simulate scienfitic processes during practical activities? Are the practical skills and problem solving skills attained during practical work transferrable to life after school? What about the role of practical work in enhancing students creativity, critical and independent thinking? In otherwords, societal expectations are changing with increasingly changing demands for new knowledge, skills and compentences (He et al., 2023; Schleicher, 2016), and as teachers, our consideration and conceptualization of the role of practical work in chemistry must be adjusted and aligned to meet these expectations.

7.2.1 Can a Curriculum Orientation Shape a Teacher's Understanding of the Place and Role of Practical Work in Science in a Given Educational Context?

The understanding behind the way a given chemistry curriculum is framed has a strong bearing on how a teacher should interpret and implement the elements of the curriculum. As described in chapter three, developers or curriculum designers lay down several reasons or explanations to justify the nature, structure and contents of the curriculum proposed. Teachers are meant in practice to accept, embrace and adopt these reasons and justifications as a starting point in transforming the curriculum contents into forms usable in classrooms. When it comes to the science curriculum, the way curriculum developers conceptualize science, its contribution and how science should be taught and learned in schools, are central elements that shape the science curriculum structure of most educational contexts, and practical work is a core element of the science curricula in different countries. For example, let us consider the statement below, extracted from the Norwegian science curriculum document:

> Students will experience science as a practical and exploratory subject. Through experience, wonder, and exploration, students will understand the world around them from a scientific perspective. By working practically and by creating their own models to solve academic challenges, students can develop creative joy, the ability to innovate and understand scientific theory. The natural sciences have a special language and subject-specific ways of thinking to explain phenomena and events. (Utdanningsdirektoratet, 2020b).

The above extract from the Norwegian science curriculum reflects an understanding where the central role and place of practical work in science draws from how the nature of science itself as well as the educational goal for science education are understood by those behind the development and structuring this science curriculum. Note that this way of looking at science and how it should be taught might not be similar in your country and the inclusion of practical work in the curriculum for your country might be justified with different reasons.

When we analyse the extract above, students are expected to experience and understand the nature of science, scientific practices and the ways of thinking and communicating science knowledge. We see here that practical work serves two purposes, that is, (i) practical work as a tool for learning science, and (ii) practical work in itself as a component of the science content. First, as a teacher, you are expected to use practical work as tool to foster students' learning of science. This is where different scholars talk about using practical work to motivate students to learn science, arouse curiosity, and interest in science explorations (Hofstein & Lunetta, 2004; Millar, 2010). Second, practical work as a component of the science content concerns students' learning about scientific practices, processes and ways of thinking. Students learn about, and perform small scale scientific practices such as preparing oxygen in the lab, synthesis of (specific) organic compounds, cracking long chain hydrocarbons into shorter isomers, extraction and isolation of antioxidants, neutralization processes, and so forth. This second purpose is largely concerned with skills training, and problem-solving abilities (Boyer, 2003; Gunstone, 1991).

The above two main purposes for practical work in science are broken down into three category goals by the science education community, that is, the cognitive, affective, and the practical goal (Hofstein et al., 2005). As a teacher, there is the expectation that you will in your pre-planning, define the intention or purpose for your intended practical work activities. In most cases and plausibly, the two purposes go hand in hand, where one purpose fills into the other. However, doing practical work for the purpose of motivating or arousing students' interest and curiosity alone could be insufficient if the motive or objective underlying the sought motivation is not realized. Moreover, there are research studies that have reported that students only get a situational motivation that lasts as long as the external driving factor lasts (Abrahams, 2009; Abrahams & Sharpe, 2010). It is thus important for teachers to consider both purposes for practical work, and not only aiming at getting their students motivated through engagement in practical actvities.

7.3 Practical Activities of Varied Degrees of Freedom and Level of Complexity

A traditional and commonly employed strategy when it comes to practical activities in chemistry classrooms is the closed-ended teacher or worksheet driven strategy. This genre of practical work has been labelled as "cookbook", recipe-type, verification-based, or a confirmatory strategy (see; Akuma & Callaghan, 2019b; Olufsen et al., 2015; Sadeh & Zion, 2012). It is a strategy that we all have at one time experienced (or still experience), either as students or teachers. Normally, students are provided with a specific (written or verbal) procedure which they follow strictly to derive at the intended aim of the activity. In some instances, when the teacher is not able to or when the conditions do not allow for a student-driven activity, the teacher him/herself carries out the activity (experiment) in the form of a demostration while students observe and take notes. This closed-ended approach to practical work activities has received a huge amount of criticism in the past (see; Anderson, 2007; Berry et al., 2001; Hofstein et al., 2004; McComas, 2005) despite having dominated the way chemistry practical work activities are carried out in schools over several decades. Among the arguments against the approach, some scholars express that students are not challenged to think about the purpose of the experiment, and that the practice does not reflect the way scientists work (McComas, 2005; Olufsen et al., 2015). Other scholars describe this practical work strategy as only adequate for developing skills such as observation, data collection and organization, and constructing inferences (Sadeh & Zion, 2012).

On the other hand, the open-ended approach to practical work activities in science has been described as enabling students to do science in ways that reflect (to a reasobale extent) the way scientists work towards generating scietific knowledge (e.g.; Akuma & Callaghan, 2019a; Lee & Krapfl, 2002; Ottander & Grelsson, 2006). The strategy considers a constructivists' learning view and it is understood to combine an inquiry-based teaching approach with other structured or unstructured teaching strategies. Students are expected to be given opportunities to manipulate equipment, materials and or other kinds of data or information, to be able to construct their understanding of the phenomena and its related scientific concepts. While this approach is widely and evidently supported (e.g.; Barnea et al., 2010; Berg et al., 2003), some scholars outline several challenges surrounding the use of this strategy of practical work. These challenges range from among other factors, the lack of time, teacher beliefs, pedagogical problems, large class sizes, safety issues, fears of abbetting students' misconceptions, assessment issues, etc. (e.g.; Cheung, 2008; Gunstone, 1991; Wang et al., 2022). To counteract these challenges, some scholars suggest that teachers should employ guided inquiry rather than open-ended inquiry and that they should allow students more time to interact, reflect and make oral presentations (e.g.; Cheung, 2008; Hofstein et al., 2005).

7.3.1 What Do These Understandings Mean to the Teacher?

From a teacher's perspective, both strategies for practical work as described above are relevant and can serve varied purposes geared towards achieving specific learning targets. There is the time or situation when following pre-specified instructions is a goal of the learning in itself. Students need to develop skills in reading, interpreting, and in following pre-specified procedures to derive at specific target results. The ability to interpret and follow pre-given instructions is a vital competence that is also needed in the world outside school, for example, in the medical laboratory where any deviation from pre-stated guidelines could have life-threating consequences, and in industries where employed personel are expected to follow algorithimic routine procedures to ensure proper and successful production processes, etc. Moreover, taking into account our understanding of the nature of science and the "scientific method", it is the practice and a requirement that scientists must and should clearly describe the methodology that has led to the kind of results or conclusions they claim. This is important to allow for the falsification process as well as for a mutiple context repeat of the experiment. This widely acceptable understanding implies that strictly following prescribed procedure in seeking to duplicate or falsify reported findings is also a part of how the science community works. Consequently, the claim that cookbook experiments do not reflect how scientists work appears not to consider the secondary researcher, who attempts to follow in the footsteps of the primary researcher, yet it is through this very process that the continuity of scientifc knowledge developments has been assured.

However, the above understandings should not lead us as teachers (teacher trainees and educators) to conclude that we need to stick to cookbook experiments in chemistry. Rather, they are meant to underscore the importance of understading the reasons for why we as teachers at times might need to choose to use the cook/book strategy in one teaching situation, and then choose to use open-inquiry in another teaching situation. Indeed, science educators introduce and use the notion of "degrees of freedom" (Herron, 1971) or the "degree of complexity" (Knain & Kolstø, 2011), and consider the "guided inquiry" instructional approach (Cheung, 2008) as a way to show the different possibilities for organizning practical activities. Starting with Herron's (1971) conceptions, a chemistry teacher can possibly plan and conduct practical work activities with "degrees of freedom" varying from 0 to 3. Accordingly, with zero (0) degree of freedom, it implies that the problem given to students is well defined as well as the end result of the activity. In addition, students are provided with concrete step by step instructions to be followed. With 1 degree of freedom, both the problem to be accomplished and the procedure or method to be followed are given. With 2 degrees of freedom, students are only provided with the task (problem) and are left to figure out what procedure or method they could use to solve the problem, and then with 3 degrees of freedom, students are given neither of the three preconditions; the problem, the method, nor the expected result (Herron, 1971; Knain & Kolstø, 2011).

With "degree of complexity", Knain and Kolstø (2011) consider the complexity or level of difficulty of the topic, the teacher's role, and the knowledge target goal. Based on these, the authors describe four variations in complexity of possible practical work activities, the low, medium, medium high and high. With the low degree of complexity, the teacher is in full control of the activity and the end result is pre-defined. The teacher is specifically targeting a specific learning goal, which for example could be about learning to make accurate observations. The degree of openness of the problem and hence its complexity increases as you go from low to high. Moreover, just like with the notion of "degrees of freedom", openness here implies that students have a wider possibility to explore how they will solve the problem. However, Knain and Kolstø's high degree of complexity (level 4) defers from Herron's level with 3 degrees of freedom in that, the latter leaves students to decide which problem to address while the former defines the problem to the students, despite the problem being open and of high complexity.

Both Knain and Kolstø's (2011) and Herron's (1971) ideas as briefly described above are useful and applicable in the chemistry classrooms. What is important is the understanding that the classroom teacher has an important role to decide the suitability and adequacy of the type or nature of practical work activity in a given teaching situation. The effectiveness of practical work activities is thus a consequence of the teachers' ability to design suitable activities targeting specific learning goals. That said, as a teacher, these questions are important to ask, (i) are these respective suggestions by Herron or Knain and Kolstø possible in chemistry classrooms? (ii) Am I able to plan and carryout practical work activities with two or more degrees of freedom? If yes, what would be the purpose or goal of these activities or experiments? (iii) Are experiments or practical work activities with a high degree of complexity as described by Knain and Kolstø (2011) relevant in my chemistry classroom? If yes, what kind of experiments or pratical work activities? How? And where are they applicable and for what purpose? These and other questions once answered, could support in obtaining a defined direction regarding what decisions can be considered or taken when it comes to practical work in chemistry.

7.4 What Students Do Learn from Science Practical Activities and How Teachers Should Enhance the Attainment of Practical Work Intended Learning Aims

At the start of the chapter, four generally conceived goals of practical work in science are mentioned, that's (i) facilitate understanding of scientific concepts, (ii) arouse students' interest and motivation, (iii) train and acquire practical skills and problem solving abilities, and (iv) attain basic knowledge about the nature of science (scientific knowledge, method and processes). Several research studies within science education have sought to establish whether practical work in science really serves these goals, or whether there are tangible learning benefits associated with practical work in science (examples include; Abrahams, 2009; Abrahams & Millar, 2008; Abrahams & Reiss, 2012; Hodson, 1990, 1996; Millar & Abrahams, 2009; Osborne, 2015). Unfortunately, this question remains unresolved up to now in 2025. Empirical evidence indicates that practical work as it is conceived and practiced in science (chemistry) teaching today (July 2025), does not seem to result in or lead to the much-anticipated learning benefits. First, teachers seem to pay much attention to and hence use almost the entire practical work sessions to ensure that students execute practical procedural tasks as given by the teacher, manipulate equipment, work with chemical substances, and complete the assigned tasks, and hence teachers are lost into these procedural activities with not much time left to focus on the key learning elements of practical work (Abrahams & Millar, 2008; Abrahams & Reiss, 2012). When it comes to the question of students being motivated or aroused by the practical activities they perform, empirical research results indicate that the effects are short-lived (Abrahams, 2009). In short, there is no clear-cut or tangible indication for what learning contribution is made by practical work in science. Educators blame this partly on how practical work and its role in teaching is conceived (or understood) by both teachers and the large part of the science education community. Teachers are faulted for having a wrong understanding of the role of practical work in science. There is the general perception and the expectation that we should have practical activities included in the science (chemistry) curriculum, and most teachers seem to be obsessed with integrating practical activities in their teaching and work tirelessly to ensure that their students accomplish practical tasks smoothly and safely. What is forgotten is that learning from practical work most likely occurs when there is a good link between what students observe/perform and the underlying scientific ideas that inform the observable or tangible practical tasks (Abrahams, 2009; Osborne, 2015).

Often, arousing students' interest, curiosity and motivation and active students' participation are at the forefront whenever questions about why we should have practical work are raised. Some teachers indeed argue that we should have practical work in chemistry because chemistry is a practical subject. But what should really be the primary goal of practical work in chemistry? Is it a sufficient justification to invest huge amounts of money to set-up school laboratories only to arouse interest and motivation in science? Osborne (2015) for example points out two aims that accordingly should primarily be the focus of practical work in science. First, to provide first-hand demonstration of phenomena and second, to provide an experience of what it means to engage in the whole experience of empirical enquiry (p. 21). Let us revisit our understanding of what chemistry is as a discipline, perhaps this will help us understand and align our own conceptions and beliefs about what the primary goals for practical work in chemistry should be. The Royal chemical society defines chemistry as "the study of the composition and properties of matter and how and why it undergoes change" (Gibney, 2018 p. 31). Accordingly, the knowledge body of chemistry has been built up through centuries of observation, experiment and reasoning, and continues to be built through these and other innovative strategies, also being supported by technological advancements. Exploring to understand matter, its composition, properties, the transformations it undergoes and why, is an ongoing process and new understandings that help to shape and improve our world are continuously emerging. Three components of chemistry are visible from the above description as summarized in Fig. 7.1, that's (i) the accumulated knowledge body of chemistry (chemical concepts), (ii) the process/methods that result in the existing body of chemistry as we know it, and (iii) the chemistry in use (the applicability of the generated knowledge body of chemistry).

This way of looking at the discipline of chemistry which is advanced by the Royal chemical society -chemistry education group, gives us a good understanding of what our students need to learn about in chemistry. That is, they should not only learn chemical concepts but get an orientation into how this body of chemical concepts has been developed, and then they should go further and explore the applicability of the chemistry concepts in shaping our world. Note that in chapter one, we described and considered the latter two categories as part of the "epistemic core of chemistry" (Erduran & Kaya, 2019), a term capturing the nature

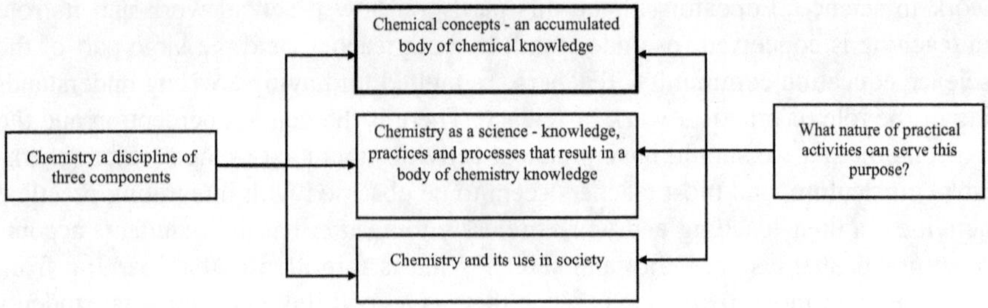

Fig. 7.1 Chemistry a discipline of three components

of chemical knowledge and practices (that's, how chemical knowledge is generated, justified, and communicated). With such a comprehensive understanding of the constituent components of the body of a knowledge of chemistry (as especially described in chapter one), it is then possible to clearly define and justify what contribution could be made by practical work in chemistry. Using Fig. 7.1, if for example the focus is on learning chemical concepts, then engaging students in practical experiments could serve well as a first-hand demonstration of phenomena as argued by Osborne (2015). Accordingly, "first-hand experience of any phenomenon that behaves in the manner predicted by the scientific account is powerful evidence of its veracity (p. 21). In other words, students' learning of chemical concepts can be reinforced when the explained chemical knowledge accounts are experienced through performance of corresponding experimental activities. A lot of teaching in schools today (2025) is more focused on students learning chemical concepts and less on the two other components shown in Fig. 7.1. This is in part due to the kind of chemistry curriculum that is being followed or implemented. The focus on reinforcing the understanding of chemical concepts is not bad either. The problem concerns rather what both teacher and students do during these practical activities, whether students have the opportunity to make the links to the scientific ideas that are the focus of the performed investigations, whether students understand the questions they are trying to solve and why, and whether there are strategies that enable students to create links between underlying chemical principles and the ongoing practical activities.

If a teacher seeks to cover Osborne's aim two, which is to provide an experience of what it means to engage in the whole experience of empirical enquiry, then s/he will need to develop and plan practical activities differently. The same will apply if the same teacher wishes to use practical work activities to promote students' understanding of the interplay between chemical knowledge and its use in society. However, planning and implementing practical work activities that target these last two components of chemistry (Fig. 7.1) can be very challenging to teachers. Science teaching today (July 2025) is highly driven majorly by participationist views about learning for example the socio-constructivists' learning views, or a combination of cognitivists and participationist views about learning (see for example; Agarkar & Brock, 2017; Chinn & Iordanou, 2023; Leach & Scott, 2003), and inquiry-based teaching approaches are highly emphasized. When it comes to practical work in science, the cookbook recipe style has been highly criticized in favor of more open-ended and student-led practical work activities. Again here, there is the danger of teachers blindly following research recommendations, without critically thinking about what the real purpose of planning and carrying out open-ended practical activities could be versus the purpose for carrying out cookbook recipe type experiments. It then goes back to you the teacher. Why are you doing what you are doing or what you intend to do? Are you planning a cookbook recipe style practical experiment because you want your students to attain or understand certain chemical concepts through experimentation, attain some skills for manipulating equipment, measurement skills, mathematical skills for computation of errors, or your goal is to help your students legitimize chemical knowledge by carrying out an experiment that shows similar results as predicted by chemical principles (theories)? Would these same experiments be possible to plan in a slightly different way to allow students more room to explore the problems without having all the information readily available to them? Will this opening-up lead to the same or even better anticipated learning results as they would for the closed-cookbook style experiments? If yes, what would you require as a teacher to be able to plan these experiments in a different way? These are some of the questions that a teacher who seeks to navigate between more closed and open-ended practical work activities should ask themselves during the planning phase of practical activities. Indeed, the productivity of any practical activity in science will rely heavily on the teacher's ability to pose these questions to him/herself and continuously seeking to find answers to them. In so doing, the teacher moves away from planning and carrying out practical work activities as culturally integrated routine practices, but as consciously planned practices in science classrooms.

While the drive towards inquiry-based practical work would offer better opportunities to students to learn science while actively involved and to engage in and experience the process of scientific enquiry, the practice is often superficially conducted by teachers in schools (for those that make the effort). In the school science curricula for a number of different countries, they have as one of the aims to promote students' understanding of scientific practices and ways of thinking in science. Without doubt, this aim could largely be attained through well-planned and executed practical work activities. Osborne (2015) critically looks at what elements could go into enabling students' experiencing of and understanding of scientific practices and ways of thinking in science. That is, (i) exploring and identifying the problem or the gap that exists, (ii) engage in proposing, developing, designing or identifying a plausible way to solve the identified gap/problem. This could involve deciding upon what kind of data will help solve the problem/gap and then engage in designing appropriate experiments that yield the relevant data, and (iii) engage in data collection, analysis and interpretation processes to produce results that will lead to inferences about the problem/gap. It is clear that teachers, even if they had sufficient competence, cannot ideally offer this experience to their students through practical work activities. They simply do not have enough time. Moreover, this process is not as straight forward as the numbering above and it is important to underscore that scientific ideas do not readily or simply emerge from a straightforward experimentation and observation process. That is why the teacher's knowledge of the epistemic practices in chemistry (Erduran & Kaya, 2019) as discussed in chapter one, is an important factor if the aim of promoting students' understanding of scientific practices and ways of thinking in science is to be achieved as a teaching goal.

If teachers do not have enough time for practical sessions that deeply engage their students to explore and understand the above aspects that reflect scientific practices and ways of thinking in science, then what options do they have? Empirical research on the effectiveness of practical work indicates that teachers fail to include activities or strategies to enable their students make links between the observations or activities performed during practical sessions and the scientific ideas that the practical activities seek to illuminate (Abrahams & Millar, 2008; Abrahams & Reiss, 2012; Osborne, 2015). The problem is not much the question of practical work activities being predominantly of a cookbook recipe style, and neither is it the teacher's failure to plan and execute more open-ended practical activities. Rather, the problem concerns the ability to consciously plan practical activities whether closed or open-ended, with clearly defined goals (aims or intentions), and with a proper strategy or set of strategies that will ensure the attainment of these target goals. Conceived this way, it means that the productivity or usefulness in terms of learning, of any type or genre of practical work activity is largely dependent on the teacher's competence and his/her knowledge of what students should learn from practical work activities.

Consequently, teachers do not have to go any further than it would be necessary to make practical work activities more productive. Moreover, strong research evidence shows that understanding science ideas takes shape through talking, reading and writing (e.g.; Norris & Phillips, 2003; Osborne, 2015; Pearson et al., 2010). Teachers can increase the effectiveness of practical work activities through strengthening the other components surrounding these activities, that's (i) the talking that goes on before, during and after hands-on-activities, (ii) the reading that goes on before- during and after hands-on-activities, and (iii) the writing that goes on before, during and after hands-on-activities. Accordingly, these three activities—call them literacy/language or communication skills (talking, reading and writing) are constitutive parts of the science itself (Norris & Phillips, 2003). They are the enabling tools with which the teacher can support students to make links between the observations or activities performed during practical sessions and the scientific ideas that the practical activities seek to illuminate. Instead of being overly occupied with fast-tracking the completion of the set practical work activities, a teacher can take a step back and think about the goal of the entire practical session before it happens and work out a program that incorporates the three literacy activities as core activities to the success of the planned practical work session.

7.5 Making Practical Work Productive: What Knowledge, Skills or Competences Are Needed?

Now that we have gone through the theoretical conceptions concerning what constitutes practical work, its place, and role in science (chemistry) classrooms, we will turn our attention to the important preparations that we teachers must make in order to be ready to exploit the potential benefits of including practical work in teaching. Teachers' knowledge and competence requirements for effective chemistry teaching were discussed in chapter two. In this part, we want to focus on teacher knowledge and skills required to plan, organize and implement productive practical work activities. Some scholars have used the term teacher's "practical work pedagogical content knowledge"—PPCK to characterize this kind of teacher knowledge (e.g.; Wei et al., 2019). For chemistry or science teachers, PPCK also constitutes teachers having sufficient laboratory related technical skills, and knowledge about chemical handling (treatment), assessment of risks related to practical work activities, and disposal of chemical waste.

7.5.1 The Teacher's Practical Work Pedagogical Content Knowledge (PPCK)

By making a few modifications to the original conceptions constituting the teacher's pedagogical content knowledge (PCK) (see; Barnett & Hodson, 2001; Kayima, 2022; van Driel et al., 2001, 2014), science educators (e,g; Chen & Chen, 2021; Wei & Liu, 2018; Wei et al., 2019), list up to five components considered to constitute the teacher's practical work pedagogical content knowledge, including,

(i) knowledge and beliefs about goals toward practical work, referring to the reasons why teachers should engage students in practical work, the orientations of their teaching, and the objectives they would like to attain with practical activities
(ii) knowledge and beliefs about the curriculum and teaching materials of practical work, mainly referring to the amount, arrangement, and features of the laboratory teaching activities in the curriculum
(iii) knowledge and beliefs about students' difficulties understanding specific experiments and activities, referring to teachers' knowledge of students' difficulties in understanding specific principles, concepts, or operations related to laboratorial activities
(iv) knowledge and beliefs about instructional strategies for teaching with practical work, referring to the specific procedures and strategies for carrying out instruction in laboratory or outside the laboratory (classroom) whenever this is possible
(v) knowledge and beliefs about assessment in teaching with practical work, involving specific standards and methods of assessing students' learning achievements in laboratory.

First, it is important that the teacher has a full understading of the purpose or role practical work serves in the teaching. This is well-addressed in the first section of this chapter. What is important to stress here is that the general goals for practical work in science as known from liturature and as also discussed in the previous section must be broken down and contextualized to individual teaching situations, and teachers should have the competence to do so. By breaking down and contextualizing, I imply re-defining the goals to suit the contextual needs at the time or the specific competence goal in the chemistry curriculum. This could also require creating small sub-goals for the intended practical work activities.

Note that the contributions of practical work activities in terms of students' learning achievement are not easily directly measurable and or quantifiable, apart from observing students' activity (verbal and non-verbal behavior) during completion of assigned practical tasks. Moreover, it is not suprising to find that the question of whether or not practical work contributes to students' learning has been repeatedly explored in research over the last 5 decades, with researchers failing to substantiate the positive learning effects of practical work (for a historical review see: Gilbert et al., 2006; Nakhleh et al., 2003). A positive correlation between practical work and students achievement is inferred when we choose to look at or investigate these effects qualitatively rather than focusing exclusively on quantitative results. This implies studying and analyzing the classroom discourse, the interaction between teacher and students as they each accomplish their roles during practical activities. In otherwords, the positive benefits of practical work are largely dependent on what the teacher actually does, what learning opportunities does s/he offer to students during practical activities, how s/he plans and executes the planned practical work activities, including how the teacher facilitates the attainment of the intended learning goal(s). The teacher thus, drawing on the relevant competance goals in the curriculum, should be able to conceptualize and develop his/her own understanding of the general goals for practical work as given in science literature in order to operationalize, use or integrate them in his/her teaching. Once a full understanding of the relevance of including practical work in chemistry teaching is realized, it becomes easy for the teacher to design different types of activities ranging from closed-ended to open-ended, since this process is dependent on the teacher's set specific learning goal(s). Moreover, the teacher's practical work pedagogical knowledge (PPCK) also includes the ability to interpret the currculum which is the guiding framework for deciding the relevant practical activitities for a given level of students.

Second, with a good mastery of chemistry content knowledge, and with the knowledge of students, teachers are expected to adapt the planned practical activities to the situational needs of their students in a given learning context. A general expectation for us chemistry teachers is that we have sufficient understanding of the degree of complexity or abstractness of the different chemistry concepts from the learner's perspective (though this is not normally the case), at least through our experience either as former chemistry students or as teachers. With this expectation, it should be possible to predict what aspects of the planned practical activities would present difficulties to students and then go ahead to plan how to support students in the process. Moreover, just like for the general subject teaching, we are also expected to be able to set teaching goals for the planned practical activities, prepare, organize, sequence, and implement the activities in a logical way, manage time schedules, as well as plan and implement an assessment strategy for evaluating the achievement of intended learning goals. This statement does not only apply to closed-ended practical activities that are mainly conducted in closed rooms, but includes all the possible practical work genres as I have discussed them in the sections before and partly in chapter nine (9).

Finally, we get to the question of how chemistry teachers come to develop this set of knowledge, competences, and attributes that constitutes the teacher's practical work pedagogical content knowledge. Well, this same or similar question has been explored in educational research by a number of educators (see for example; Appleton, 2006; Barnett & Hodson, 2001; Chen & Chen, 2021; Fan, 2014; Grossman, 1990; Kayima, 2022; Marco-Bujosa et al., 2017). Sources through which practical work pedagogical content knowledge is acquired as reported in literature include evolving experiential knowledge and skills gained over the years as a student in the primary and secondary school, through didactics courses during teacher training, in-service training experiences such as through workshops, conferences, seminars, classroom peer-observations—learning from experienced colleagues, through reading science education literature and the use of online resources, studying curricula materials, and through school practice for teacher-trainees. Regardless of which source is considered, what is important for us as chemistry teachers is to have the awareness that we need more than a chemistry content knowledge to be able to teach our students effectively. This awareness is an important step for us to start to search for ways through which we can acquire those skills and competences that we need to do our job.

7.5.2 The Teachers' Need for Laboratory Skills and Knowledge

A set of technical knowledge and skills related to practical work activities is an important component of a teacher's knowledge base that is needed for incorporating practical work in the teaching. This set of knowledge and skills includes but not limited to,

i. knowledge about specifications for selected chemicals to be used in classrooms, including knowledge about the health, safety, and disposal guidlines,
ii. instrumentation knowledge and skills—knowledge about the different relevant laboratory equipment and how they are safely used in the lab or during experiments,
iii. up-to date knowledge about the more-safe and evironmentally friendly chemicals as possible replacements for the non-environmental and convetional

chemicals: includes knowledge about the possible sources for cost effective experiments applicable in classroom settings—(green and sustainable chemistry—refer to Chapter 10)

iv. the ability to work with highly concentrated solutions, determine what concentration levels should be safe and appropriate for the students to work with, and being able to prepare these in the lab.

v. relevant knowledge or skills for stocktaking and handling of expired chemicals. This includes knowledge about the properties of different chemicals that are part of the chemistry curriculum and taking into account the radioactivity levels, UV light exposure, etc.

First, the teacher's technical knowledge and skills are central especially when the planned activities involve students working with chemicals. Normally, your students and the entire school leadership have put their trust in you the chemistry teacher, and it's your resposibility to ensure that students will be safe when working in the lab, manuplating equipment, and handling chemicals. The same applies when students are working from outside the lab (could be field-work) as long as the activities involve treatment of chemicals. Moreover, prior to administering any tasks involving handling chemicals where students will work independently, a risk assesmment is expected to be carried out by you the teacher. You are expected to make the necessary safety preparations, including orienting students about disposal procedures to be followed based on the nature of chemicals being used, as well as measures to ensure first aid and exit measures in case of any unexpected accidents.

The underlying assumption is that, once employed as a chemistry school teacher, your terms of reference include the responsibilty of managing the chemistry lab unit and so is the freedom to decide which practical activities to include in your teaching as long as these are within the curriculum framework. Research has shown that most teachers especially the novice ones, feel very insecure when it comes to planning and including practical activities in their teaching (e.g.; Cheung, 2008; Davis et al., 2006). This in part is due to the lack of the relevant technical knowledge and skills, despite that the teacher might have sufficient knowledge of chemistry. For example, when schools are to procure chemicals for use in the lab, in most cases the purchase is made in bulk. Most of the chemicals used in classrooms such as mineral acids and bases are often highly concretrated when they are bought into the school lab (see Fig. 7.2 for example). The expectation is that you the chemistry teacher, will determine what concentration levels are safe for students and then go ahead to prepare these and make them available for use in your teaching.

The sample task below represents some of the teacher's preparatory work and responsibilities we have discussed above. Read through the task and reflect on how you would approach the task and what this implies in terms of the laboratory knowledge and skills you need as a chemistry teacher.

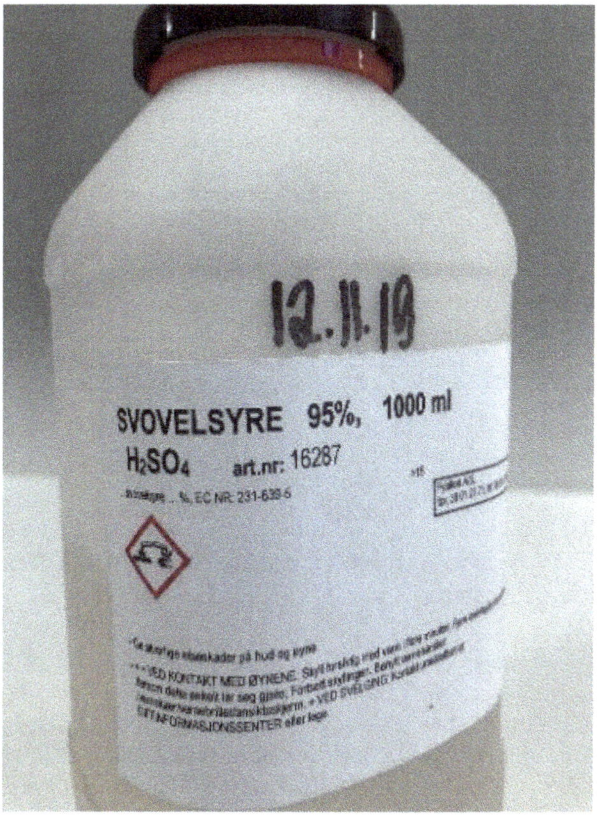

Fig. 7.2 A bottle containing a 95% Sulphuric acid solution

A newly qualified chemistry teacher A has been on the job in school X for eight weeks now. Since the start of his employment, he has not been able to carry out any practical work activities or any sort of experiment with students. In the two weeks that follow, the teacher reads thoroughly through the current school curriculum for chemistry and decides to plan to conduct a practical work activity targeting the following learning aim:

> ***"students will learn about and be able to "perform volumetric and gravimetric titration analysis and discuss the use of titration analysis":*** extracted from the chemistry curriculum for upper secondary school in Norway (Utdanningsdirektoratet, 2020a)

The teacher establishes the link between the learning aim provided in the currculum and the chemistry knowledge background students will need to be able to succeed during the experiment. Thus, he uses the next four weeks to introduce aspects about the mole concept (formular mass, mole and related calculations, balancing and using chemical reactions to establish concentrations in terms of moles) etc. For the experiment of interest, teacher A decides that students should carryout a strong acid -strong base neutralization titration experiment, since according to him, this would be simple for the students to begin with. The teacher's plan is to design a semi-closed experiment, where the final result is known and predictable, but students are expected to use their previous knowledge reflections to fill in the missing parts for the experiment to succeed. Accordingly, students will be provided with 500 mls of a 0.2 Molar sulphuric acid solution sample, and will use it to establish the concentration a given sample of a strong base. Students will not be given the procedure for this experimental activity, but are expected to consult relevant sources to find the appropriate and more safer method that will enable them to succeed with the experiment. The teacher nevertheless plans to watch over the whole process closely, to ensure students' safety during the experiment.

The teacher has 90 min every Friday set for chemistry lessons as per the school teaching timetable. Two days before Friday, he goes to the science lab to find the relevant chemicals and equipment that he will need for the activity. Fortunately or unfortunately, the teacher finds no ready chemicals for use, but only empty bottles in cupboard for mineral acids, and for strong bases. In the lab, is a small room with a bigger curpboard labelled "Radioactive". The teacher opens the cupboard and inside, he finds one flask of hydrocholoric acid with 75% purity and another flask of sulphuric acid with 95% purity. The flasks are also marked with R.D 1.18 and 1.82 respectively. On the side of the cupboard, the teacher finds two flasks with solid perates of sodium hydroxide

Question: *Discuss how the teacher proceeds from here, reflect on the importance of the teacher's technical knowledge and skills*

7.6 Safety in the School Laboratory or When Working with Chemicals

Chemistry as a subject is concerned with studying the composition and properties of substances, the changes they undergo, how and why. Normally the term "chemical substance" is used when referring to substances in the context of science or chemistry, where this substance could be an element, compound or a substance mixture. Whether performed in the school science laboratory or outside in a pre-organized learning environment, chemistry practical work activities will mostly involve students and teachers working with chemicals and some specific equipment as they seek to explore and understand matter (substances), its properties and behavior. While not all chemicals are dangerous to human health and the environment (at least not when in very small amounts), working with chemicals comes with safety responsibilities on the part of the teacher, pertaining to the teacher him/herself, the students, the school laboratory room or school premises, and the environment. Fortunately, there are already principles and regulations in place to guide how different interest groups should handle and work with chemicals to ensure safety for humans (and other living organisms), property, and the environment. The United Nations (UN) international team of hazard communication experts developed a Globally Harmonized System (GHS) of Classification and Labelling of Chemicals, comprising (i) the rules for classifying hazards of chemical products and (ii) tools for communicating these chemical hazards. GHS has been adopted and contextualized by several individual countries. For example, EU member states have what they call the "Classification, Labelling and Packaging (CLP)" Regulation that is based on the United Nations' GHS, aimed at ensuring "a high level of protection of health and the environment, as well as the free movement of substances, mixtures and articles." The CLP regulation provides guidelines

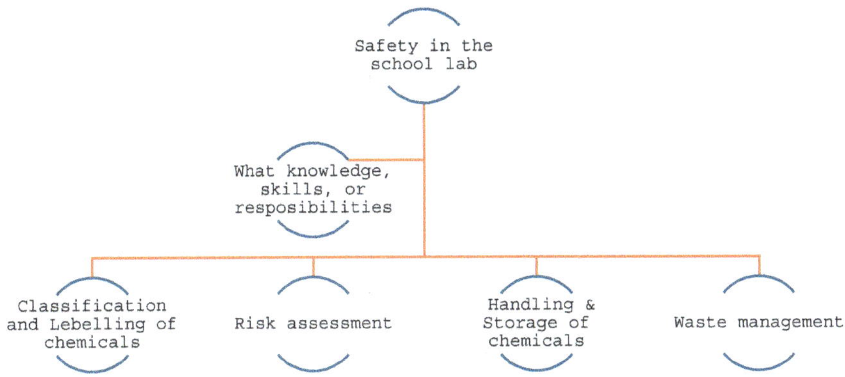

Fig. 7.3 What you need to know to ensure safety in the school laboratory

for classifying, labelling and packaging of substances or mixtures https://echa.europa.eu/regulations/clp/legislation.

Our safety responsibilities as teachers when it comes to handling and working with chemicals concern knowing how to handle different chemicals, laboratory and other relevant equipment, how to assess risks involved when handling certain chemicals, how to avoid or reduce the occurrence of accidents during laboratory activities, and how to react or deal with inevitable accidents when they occur. Thus, it is expected that you the chemistry teacher are able to (i) to read, interpret and extract information from safety data sheets (SDS) and use it to perform a risk assessment, (ii) able to perform a risk assessment before, during, and after any use of chemicals and equipment, (iii) able to follow either the UN's GHS, EU's CLP stipulated guidelines, or the guidelines adopted by your country (local district), concerning the classification, labelling, handling and storage of chemicals, and (iv) are able to sort and manage chemical waste as according to the chemical waste management guidelines (Fig. 7.3).

7.6.1 Classification and Labelling Chemicals

As mentioned earlier, the UN's GHS comprises the rules for classifying hazards of chemical products as well as the tools for communicating chemical hazards. For EU member countries, chemical substances or mixtures are labelled following the CLP regulation. You (the chemistry teacher) need to find out what kind of regulations are followed in your country if you are outside the EU. Currently (as of 2025), the key words and tools used for communicating the hazards of chemicals (labelling of substances or mixtures) include the use of signal words such as *danger*, or *warning*, pictograms, hazard statements, and precautionary statements. The hazard of a substance or mixture is described as the potential to cause harm based on the intrinsic properties of that substance or mixture. Thus, *hazard statements* as applied in the GHS regulation describe the nature of hazards of a substance or mixture (in some cases even the degree of hazard), while *precautionary statements* inform about the recommended measures to minimize or prevent adverse effects in case of exposure to a hazardous substance or mixture. The *signal words (danger or warning)* indicate the severity of the hazard, whereas *Pictograms* are a graphical composition comprising a symbol and other graphical elements meant to convey specific information on the hazard of concern. Science (chemistry) teachers are expected to understand and recognize this universal labelling and communication framework or otherwise have a good knowledge about where to find this information, which is very important for assessing risks prior to any use of chemical substances or mixtures. In Table 7.1 is a list of 9 hazard pictograms that you most likely will encounter in the science/chemistry laboratory. Table 7.2 shows an example of how chemical hazards are communicated using the mentioned keywords and tools. The information used in Table 7.2 is taken from Online-Resources-for-Chemical-Safety-in-Science-Education (2022, December 5), which are a result of a three -year ERASMUS + project (2020–2023) ran by the University of Oslo (Norway), the University of Helsinki (Finland), the University of Ljubljana (Slovenia), and Stockholm University (Sweden)—https://chesse.org/

As mentioned earlier, some chemical substances that are bought into the school laboratory for use in teaching are often in concentrations not usable in a school learning environment. Examples include stock solutions of mineral acids such as Sulphuric acid and Hydrochloric acid, whose concentrations could range between 40 and 98% purity, implying higher acid concentrations up to about 18 Molar or even higher. Figure 7.4 for example shows two solutions of mineral acids, a 12-molar concentrated solution of Hydrochloric acid and a 95%, (approx. 17.78 Molar) stock solution of Sulphuric acid. Such higher concentrations cannot be used in classrooms by students and teachers need to scale these down by preparing lower, safer and hence appropriate concentrations.

When a teacher prepares his/her own mixtures or solutions, it is required that the teacher labels these also following either the GHS or CLP or the adopted labelling guidelines for users of chemical substances or mixtures. In some cases, exemptions from the full labelling are provided in the regulations (either GHS or CLP). For example, according to CLP regulation (Annex 1, Sect. 1.5), some chemical substances or mixtures are exempt from the labelling requirement where the contents are 125 ml or less, depending on the nature of hazard for substances or mixtures of concern. You will need as a teacher to familiarize yourself with the CLP regulations (for those working in EU) or with the guidelines adopted in your country for information regarding labelling exemptions. In Table 7.2 for example, different concentrations of sodium hydroxide solutions are labelled differently using the CLP/GHS hazard communication tools. Accordingly, Sodium hydroxide solutions of concentration below 0.125 ($C < 0.5\%$) are not classified as harmful mixtures and hence exempted from the labeling requirement.

Table 7.1 Hazard Pictograms

Pictograms		
 **Explosive**	 **Acute toxicity**	 **Gas under pressure**
 **Flammable**	 **Health hazard/Hazardous to Ozone layer**	 **Corrosive**
 **Oxidizing**	 **Serious health hazard**	 **Hazardous to the environment**

7.6.2 Handling and Storage of Chemicals

Ensuring proper and safe handling of chemical substances or mixtures in the school is a responsibility shared between the school principal (head teacher in some schools) or his/her delegated person, the head of the science department, and you the chemistry teacher (or the science teacher in cases where this is the case). You are expected to work together as a team to identify all hazardous chemical substances or mixtures which have been bought into the school for use in teaching either directly or indirectly. In most current schools, the leadership has ensured that the school has a register for chemical substances and mixtures for capturing the details of all bought in chemicals, including documenting the nature of hazards associated with specific chemical substances. This chemical register is normally updated frequently as either new chemicals are bought in or when there are changes in how risks of certain chemical hazards are or should be reported.

Table 7.2 An example for use of Signal word, H-statement, P-statement and a Pictogram to communicate a chemical hazard (Online-Resources-for-Chemical-Safety-in-Science-Education, 2022, December 5)

Sodium hydroxide, NaOH				
Concentration	Hazard Pictogram	Signal Word	H-statements	P-statements
$C \geq 0.5$ M $C \geq 2\%$		Danger	Causes severe skin burns and eye damage	Wear eye protection. If SWALLOWED: Rinse mouth. Do NOT induce vomiting. IF ON SKIN (or hair): Take off immediately all contaminated clothing. Rinse skin with water (or shower). IF IN EYES: Rinse cautiously with water for several minutes. Remove contact lenses, if present and easy to do. Continue rinsing. Immediately call a doctor
0.5 M > $C \geq 0.125$ M $2\% > C \geq 0.5\%$		Warning	Causes skin irritation. Causes serious eye irritation	Wear eye protection. If in EYES: Rinse cautiously with water for several minutes. Remove contact lenses, if present and easy to do. Continue rinsing. If eye irritation persists: Get medical attention
$C < 0.125$ M $C < 0.5\%$	Not classified as a harmful mixture			

The register is also updated frequently to match the current school chemical stock (that is, during stocktaking).

One of the responsibilities of the principal (head of school) is to ensure that the school has in its possession the most current safety data sheets (SDS) from the manufacturer or supplier. A safety data sheet provides detailed and comprehensive information relating to a chemical substance's hazard, health effects due to exposure, measures to protect against the risk of exposure, and information about the emergency procedures upon exposure. Like the chemical register, SDS must be accessible to especially all people working with chemical substances or mixtures. Moreover, it is recommended that SDS are stored in places accessible to all employees in the school. Note that some chemical substances are not suitable for use in schools due to a higher risk level. Information about these substances is provided in the CLP regulation for those in EU, or in the regulation adopted by your country. The principal (head of school) is expected to obtain and make this information available to the teachers.

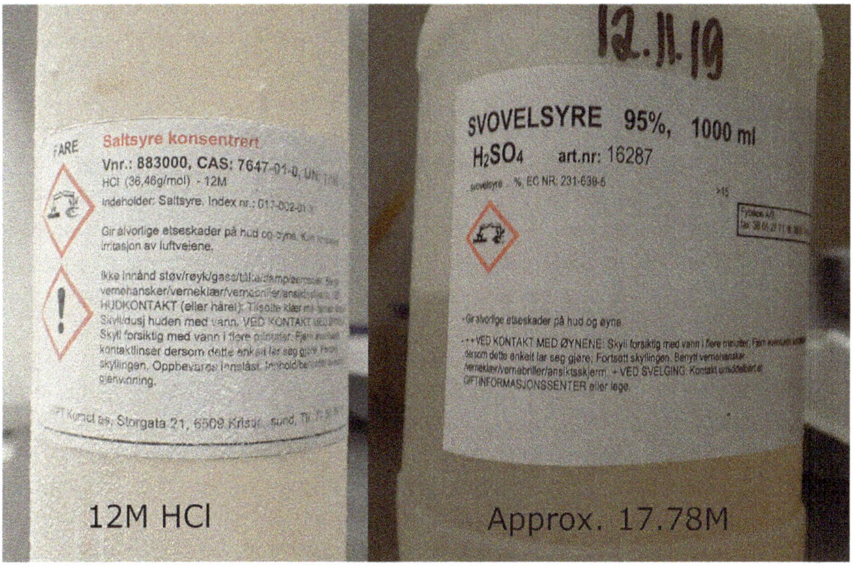

Fig. 7.4 Stock solutions of HCl (12 Molar) and H_2SO_4 (95%, approx. 17.78 Molar)—pictures taken in a science laboratory storage room at Norwegian Teacher education institution

7.6.2.1 Chemical Storage

When it comes to proper and safe storage of chemical substances and mixtures or any other hazardous substances or objects, most countries have in place what is often referred to as a "dangerous goods regulation" that provides guidelines for handling and storage of dangerous goods. The school principal is expected to find the relevant regulation or a set of guidelines relevant to school settings if this has been developed by the education department or ministry of education in your country. Nevertheless, existing chemical handling and storage guidelines will generally be the same since these are adopted from the general GHS regulation. The key important considerations include.

i. Hazardous chemicals must have sufficient distances between them to eliminate the risk of fire, explosion or the accumulation of gases or vapors from leakages of spillage.
ii. All chemicals must be stored according to classes, while considering incompatibilities within a class
iii. Chemicals prepared or taken from their original containers into other containers must be labelled according to CLP (or GHS) regulation. The used containers must be suitable for the purpose and if not, it is recommended that hazardous chemicals be stored in original and labelled containers.
iv. Hazardous chemicals are stored in cabinets that are approved by regulating authorities, however, given quantities of different chemical substances might be stored in the laboratory other than in cabinets.

v. Physical barriers should be used whenever possible or otherwise measures should be in place to prevent exposure to hazardous chemical substances. Protective gear must be readily available and must be used when handling all hazardous chemical substances or mixtures.
vi. Equipment is stored separately in cabinets away from chemicals.
vii. The chemical storage room should have safety equipment installed if it is detached from the classroom. However, if the storage room is near the classroom and with emergency equipment and no physical barrier, there is no need for additional safety equipment in the storage room. The location of chemical storage areas (rooms) must be chosen in consideration of safe transportation of chemicals.

Table 7.3 shows an overview of the different chemical classes and divisions along with some hints regarding storage. Please note that this table is incomplete and does not give detailed information. You must refer to SDS for proper storage procedures. It is important that you have full information regarding the compatibility and incompatibility of chemical substances and mixtures. If there is no option but to store incompatible chemicals in the same cabinet, then those responsible must ensure that there is a physical segregation (secondary containment).

7.6.3 Risk Assessment

Science or chemistry teachers intending to carry out practical activities where chemical substances or mixtures and any equipment with potential chemical, physical or safety hazards are used, must carry out a risk assessment before, during, and after the activities. Risk assessment involves identifying the hazards associated with concerned chemical substances, mixtures or equipment to be used and enacting control measures that reduce those identified risks. Performing a risk assessment when handling or working with hazardous chemical substances and equipment bears legal responsibility as per the GHS or CLP regulation. The principal of the school therefore must ensure that information that will support teachers in carrying out good risk assessment practices such as SDS is readily available and accessible to teachers. Note that hazards could come from the very chemicals being used, the equipment or the action process itself when students or teacher are working or manipulating these items. Moreover, physical hazards could be conditions that might cause harm, for example when compressed gas or steam is released, unintended contact with hot objects, run-away chemical reactions, noise, sharp objects, surface issues such as a wet floor, unintended obstructions, fire, and electricity. Safety hazards on the other hand account for other unsafe conditions that could result in harm to people and property such as students' hair or clothing that gets entangled into equipment and thereby causing spills or fire, could be equipment malfunctioning, or students not paying attention to laboratory rules and obligations, among other things. During risk assessment therefore, the teacher should have concrete response-actions to questions such as (i) what could

Table 7.3 Different chemical classes and divisions along with storage hints

Class	Division	Storage/recommendation
1: Explosives	1.1 substances with a mass explosion hazard	Highly flammable substances or mixtures must be stored in tightly closed bottles in a fireproof cabinet with mechanical ventilation, in a secure location away from other chemicals Must be stored at least 5 m away from any potential source of ignition or heat if not stored in specialized chemical storage cabinet
	1.2 have a projection hazard but not a mass explosion hazard	
	1.3 have a fire hazard and either a minor blast hazard or a minor projection hazard or both, but not a mass explosion hazard	
	1.4 present no significant hazard	
	1.5 very insensitive substances with a mass explosion hazard	
	1.6 extremely insensitive articles with no mass explosion hazard	
2: Gases	2.1 Flammable gases	Must be stored indoors except for aerosols and compressed air
	2.2 Non-flammable, non-toxic gases	
	2.3 Toxic gases	
3: Flammable liquids		Store in flammable storage cabinet (in excess of 40 L) within secondary containment
4: Flammable solids	4.1 Self-reactive substances and solid desensitized explosives	
	4.2 Substances liable to spontaneous combustion	
	4.3 Substances which in contact with water emit flammable gases	To be stored in a moisture proof cupboard or dangerous goods cabinet without plumbing mixtures or any liquids
5: Oxidizing substances and Organic peroxides	5.1 Oxidizing substances	Stored in secondary containment and separate from flammable and combustible materials

(continued)

Table 7.3 (continued)

Class	Division	Storage/recommendation
	5.2 Organic peroxides	Highly reactive and flammable – must be stored in a cool place, at least 3 m from all others dangerous goods and shielded from all heat sources
6: Toxic and infectious substances	6.1 Toxic substances	Potassium and Sodium cyanides are extremely toxic and should be removed from use in schools. Inorganic cyanides (if available) must be kept under lock and key, and at least 5 m away from any acids
	6.2 Infectious substances	
7: Radioactive Material		Must be stored in separate, well-labelled containers or compartments within a locked metal container. The locked metal container must be permanently labelled to indicate that it contains radioactive substances
8: Corrosive substances		Concentrated strong acids and bases/alkalis should be stored at least 1.5 m apart or otherwise separated by an impervious barrier and provided with separate spill retention Acids must be at least 5 m away from any inorganic cyanides if there is any in the chemical storage room

go wrong, (ii) what can be done in case it happens? and (iii) what feasible steps reduce the risk that it should happen?

There are several precautionary measures that teachers can take to reduce the risks associated with the use of chemical substances. One of them is to scale down quantities and concentrations of solutions. Scaling down quantities or concentration levels also aims at teachers designing activities or experiments which have a reduced demand for personal protective equipment. Moreover, a teacher's knowledge about the best alternative activities or if not, an awareness about less hazardous chemicals that could be used to achieve the same target learning goal is very important for promoting and ensuring safety in the lab. Drawing on and applying green chemistry principles (Anastas, 2003; Anastas & Warner, 1998; Idul &

Walag, 2024), and working towards replacing traditional experiments with green alternatives is encouraged whenever possible. This involves for example using safer—less toxic or corrosive chemicals, being less wasteful and environmentally friendly when choosing which chemicals to use. Nevertheless, installed fume cupboards or fume hoods should be used, especially when the experiments involve the use of large quantities of chemical substances or when students are working in big groups. Fume cupboards can limit exposure to hazardous fumes, vapors or dust.

Routine practices for both teacher and students are also important for ensuring safety. For example, handling of glassware with care, use of protection gear such as goggles, lab-coats, putting on clothed shoes while in the lab, use of hair clips for those with long hair, avoiding eating or drinking while in the laboratory (better to excuse oneself for a moment in case one cannot wait), ensuring that students know how to use lab equipment that are more prone to causing accidents such as how to use a Bunsen burner, etc. The teacher is expected to take students through these important laboratory practices and to cultivate a practice culture that is informed by these and other safety related practices. Moreover, beyond these precautionary measures, risk assessment involves establishing what actions to take in the event an accident happens. The questions the teacher must answer here include establishing what kind of first aid is possible under such circumstances, whether there is a direct telephone line to first aid services, whether the fire alarms are working as they should, or whether students have full knowledge of the available exit routes and hallways in case such a need should arise.

7.6.4 Chemical Waste Management

Chemistry or science teachers who plan and use chemical substances or mixtures in their teaching must also ensure that the waste that comes from these are responsibly managed and disposed according to the waste management guidelines in place (could be from CLP or GHS regulation or locally adopted chemical waste management guidelines). Again here, being able to extract information from SDS is very important for the teacher to know the nature of hazard for a given chemical substance and how the chemical should be handled.

An ERASMUS + project conducted between 2020 and 2023 by a team of educators from University of Oslo (Norway), University of Helsinki (Finland), University of Ljubljana (Slovenia), and Stockholm University (Sweden), resulted in well-developed and useful online resources for chemical safety in science education (Online-Resources-for-Chemical-Safety-in-Science-Education, n.d). These resources are open to all, at least for those within the EU, and include aspects about the teachers' responsibilities, routines and training, risk assessment and substitution, labelling, storage and waste management, including other useful information. Regarding chemical waste management, the ERASMUS + team advances a waste management regime that follows EU's Waste Framework Directive (https://environment.ec.europa.eu/topics/waste-and-recycling/waste-framework-directive_en) and encourages the teacher to (i) reduce

Table 7.4 General chemical waste management guidelines

Chemicals/mixtures that can go in a regular trash/bin	Chemicals/mixtures that can go down the drain	Chemicals/mixtures that cannot go down the drain	Hazardous wastes
Nonradioactive Not a biological hazard Not flammable, reactive, corrosive, toxic, or not listed as hazardous waste by EPA	All those meeting the criterion for regular disposal in a trash/ cabin Acids and base solutions og pH between 5.5 and 10.5 Combinations of cations and anions (should not be strong acids or bases)	There are several, check SDS Examples: Insoluble solids Oil or grease Water- soluble polymers Explosives Flammables immiscible in water Nitro compounds Toxic chemicals Substances with a boiling point lower than 50 °C	Heavy metal solutions— Organic solvents Corrosive liquids

or avoid waste and hazards, (ii) replace hazardous substances with less or non-hazardous substitutes, (iii) use chemical knowledge and skills to make waste less hazardous (e.g., neutralize, dilute, evaporate, precipitate, filter, decompose, etc.), and (iv) carefully follow waste storage and disposal guidelines for hazardous waste that cannot be made less hazardous, and whose inevitable use cannot be avoided (Online-Resources-for-Chemical-Safety-in-Science-Education, 2023, January 3). For those with access, this digital resource could be useful for you when planning how to manage and work with chemical substances and mixtures at your local school.

If for any reason, you are not able to access digitally available resources on working with and managing chemical substances as a teacher, this section on safety in the school lab is written to provide you with the basic knowledge and the relevant information with which you can safely work in the school lab, while observing your safety responsibilities. As a general principle, we have (i) chemical waste that could be directly and readily put into a trash can/bin, (ii) chemical waste that can be washed into the sink and down the drain system, (iii) chemical waste that must not go down the drain and we cannot wash this into the sink, and (iv) then we have hazardous wastes. In Table 7.4, I have given some of the examples under each category. Remember, and as already mentioned, information about the nature of hazard and how a certain chemical substance or mixture is handled is provided in the SDS.

References

Abrahams, I. (2009). Does Practical Work Really Motivate? A study of the affective value of practical work in secondary school science. *International Journal of Science Education, 31*(17), 2335–2353. https://doi.org/10.1080/09500690802342836

Abrahams, I. (2011). *Practical work in secondary science: A minds-on approach.* Continuum International Publishing Group.

Abrahams, I., & Millar, R. (2008). Does Practical work really work? A study of the effectiveness of practical work as a teaching and learning method in school science. *International Journal of Science Education, 30*(14), 1945–1969. https://doi.org/10.1080/09500690701749305

Abrahams, I., & Reiss, M. J. (2012). Practical work: Its effectiveness in primary and secondary schools in England. *Journal of Research in Science Teaching, 49*(8), 1035–1055. https://doi.org/10.1002/tea.21036

Abrahams, I., & Sharpe, R. (2010). Untangling what teachers mean by the motivational value of practical work. *School Science Review, 92*(339), 111–115.

Agarkar, S., & Brock, R. (2017). Learning theories in science education. In K. S. Taber & B. Akpan (Eds.), *Science education: An international course companion* (pp. 93–103). SensePublishers. https://doi.org/10.1007/978-94-6300-749-8_7

Akuma, F. V., & Callaghan, R. (2019a). Characterising extrinsic challenges linked to the design and implementation of inquiry-based practical work. *Research in Science Education, 49*(6), 1677–1706. https://doi.org/10.1007/s11165-017-9671-x

Akuma, F. V., & Callaghan, R. (2019b). A systematic review characterizing and clarifying intrinsic teaching challenges linked to inquiry-based practical work. *Journal of Research in Science Teaching, 56*(5), 619–648. https://doi.org/10.1002/tea.21516

Anastas, P. T. (2003). *Meeting the challenges to sustainability through green chemistry.* Green Chemistry.

Anastas, P. T., & Warner, J. C. (1998). *Green chemistry: Theory and practice.* Oxford University Press.

Anderson, R. D. (2007). Inquiry as an organizing theme for science curricula. In S. K. Abell, K. Appleton, & D. Hanuscin (Eds.), *Handbook of research on science education.* Routledge.

Appleton, K. (2006). Science pedagogical content knowledge and elementary school teachers. In K. Appleton (Ed.), *Elementary science teacher education: International perspectives on contemporary issues and practice* (pp. 31–54). Lawrence Erlbaum Associates.

Barnea, N., Dori, Y. J., & Hofstein, A. (2010). *Development and implementation of inquiry-based and computerized-based laboratories: reforming high school chemistry in Israel* (pp. 218–228).

Barnett, J., & Hodson, D. (2001). Pedagogical context knowledge: Toward a fuller understanding of what good science teachers know. *Science Education, 85*(4), 426–453. https://doi.org/10.1002/sce.1017

Berg, C. A. R., Bergendahl, V. C. B., Lundberg, B., & Tibell, L. (2003). Benefiting from an open-ended experiment? A comparison of attitudes to, and outcomes of, an expository versus an open-inquiry version of the same experiment. *International Journal of Science Education, 25*(3), 351–372. https://doi.org/10.1080/09500690210145738

Berry, A., Gunstone, R., Loughran, J., & Mulhall, P. (2001). Using laboratory work for purposeful learning about the practice of science. In H. Behrendt, H. Dahncke, R. Duit, W. Gräber, M. Komorek, A. Kross, & P. Reiska (Eds.), *Research in science education—past, present, and future* (pp. 313–318). Springer Netherlands. https://doi.org/10.1007/0-306-47639-8_44

Boyer, R. (2003). Concepts and skills in the biochemistry/molecular biology lab. *Biochemistry and Molecular Biology Education, 31*(2), 102–105. https://doi.org/10.1002/bmb.2003.494031020192

Chen, B., & Chen, L. J. (2021). Examining the sources of high school chemistry teachers' practical knowledge of teaching with practical work: From the teachers' perspective. *Chemistry Education Research and Practice, 22*(2), 476–485. https://doi.org/10.1039/d0rp00246a

Cheung, D. (2008). Facilitating chemistry teachers to implement inquiry-based laboratory work. *International Journal of Science and Mathematics Education, 6*(1), 107–130. https://doi.org/10.1007/s10763-007-9102-y

Chinn, C. A., & Iordanou, K. (2023). Theories of learning. In N. G. Lederman, D. L. Zeidler, & J. S. Lederma (Eds.), *Handbook of research on science education* (pp. 89–120). Routledge Taylor & Francis Group.

Cooper, M. M., & Stowe, R. L. (2018). Chemistry education research—From personal empiricism to evidence, theory, and informed practice. *Chemical Reviews, 118*(12), 6053–6087. https://doi.org/10.1021/acs.chemrev.8b00020

Davis, E. A., Petish, D., & Smithey, J. (2006). Challenges new science teachers face. *Review of Educational Research, 76*(4), 607–651. https://doi.org/10.3102/00346543076004607

Dewey, J. (1929). *The quest for certainty.* Southern Illinois University Press.

El Kharki, K., Bensamka, F., & Berrada, K. (2020). Enhancing practical work in physics using virtual javascript simulation and LMS platform. In D. Burgos (Ed.), *Radical solutions and eLearning: Practical innovations and online educational technology* (pp. 131–146). Springer Singapore. https://doi.org/10.1007/978-981-15-4952-6_9

Erduran, S., & Kaya, E. (2019). *Transforming teacher education through the epistemic core of chemistry.* Springer.

Fan, L. (2014). *Investigating the pedagogy of mathematics: How do teachers develop their knowledge?* Imperial College Press.

Gibney, D. (2018). Towards an ideal chemistry curriculum. *School Science Review, 100*(370), 30–35.

Gilbert, J. K., De Jong, O., Justi, R., Treagust, D. F., & Van Driel, J. H. (2006). *Chemical education: Towards research-based practice* (Vol. 17). Springer Science & Business Media.

Gott, R., & Duggan, S. (1996). Practical work: Its role in the understanding of evidence in science. *International Journal of Science Education, 18*(7), 791–805. https://doi.org/10.1080/0950069960180705

Grossman, P. L. (1990). *The making of a teacher: Teacher knowledge and teacher education.* Teachers College Press.

Gunstone, R. F. (1991). Reconstructing theory from practical experience. In B. Woolnough (Ed.), *Practical science* (pp. 67–77). Open University Press.

He, P., Krajcik, J., & Schneider, B. (2023). Transforming standards into classrooms for knowledge-in-use: An effective and coherent project-based learning system. *Disciplinary and Interdisciplinary Science Education Research, 5*(1), 22. https://doi.org/10.1186/s43031-023-00088-z

Herron, M. D. (1971). The nature of scientific enquiry. *The School Review*, *79*(2), 171–212. http://www.jstor.org/stable/1084259

Hodson, D. (1990). A critical look at practical work in school science. *School Science Review, 71*(256), 33–40.

Hodson, D. (1996). Practical work in school science: Exploring some directions for change. *International Journal of Science Education, 18*(7), 755–759. https://doi.org/10.1080/0950069960180702

Hodson, D. (1998). Mini-special issue: Taking practical work beyond the laboratory. *International Journal of Science Education, 20*(6), 629–632. https://doi.org/10.1080/0950069980200601

Hofstein, A., & Lunetta, V. N. (2004). The laboratory in science education: Foundations for the twenty-first century. *Science Education, 88*(1), 28–54. https://doi.org/10.1002/sce.10106

Hofstein, A., Navon, O., Kipnis, M., & Mamlok-Naaman, R. (2005). Developing students' ability to ask more and better questions resulting from inquiry-type chemistry laboratories. *Journal of Research in Science Teaching, 42*(7), 791–806. https://doi.org/10.1002/tea.20072

Hofstein, A., Shore, R., & Kipnis, M. (2004). Providing high school chemistry students with opportunities to develop learning skills in an inquiry-type laboratory: A case study. *International Journal of Science Education, 26*(1), 47–62. https://doi.org/10.1080/0950069032000070342

Idul, J. J. A., & Walag, A. M. P. (2024). Integrating green chemistry and sustainability principles to a secondary science curriculum: A mixed-methods needs assessment. *Journal of Chemical Education, 101*(7), 2765–2778. https://doi.org/10.1021/acs.jchemed.4c00341

John-Steiner, V., & Mahn, H. (1996). Sociocultural approaches to learning and development: A Vygotskian framework. *Educational Psychologist, 31*(3/4), 191. https://doi.org/10.1080/00461520.1996.9653266

Johnstone, A. H. (1982). Macro and microchemistry. *School Science Review, 64*(227), 377–379.

Johnstone, A. H. (1991). Why is science difficult to learn? Things are seldom what they seem. *Journal of Computer Assisted Learning, 7*(2), 75–83.

Kayima, F. (2022). The role of reflective practice in mediating development of pre-service science teachers' professional and classroom knowledge. *Interdisciplinary Journal of Environmental and Science Education, 18*(1), e2262. https://doi.org/10.21601/ijese/11364

Kidman, G. (2012). Australia at the crossroads: A review of school science practical work. *Eurasia Journal of Mathematics Science and Technology Education, 8*(1), 35–47. <Go to ISI>://WOS:000313643100005

Knain, E., & Kolstø, S. D. (2011). *Elever som forskere i naturfag*. Universitetsforlag.

Lazarowitz, R., & Tamir, P. (1994). Research on using laboratory instruction in science. In D. L. Gabel (Ed.), *Handbook of research on science teaching and learning* (Vol. 1994, pp. 94–130). Macmillan.

Leach, J., & Scott, P. (2003). Individual and sociocultural views of learning in science education. *Science & Education, 12*(1), 91–113. https://doi.org/10.1023/A:1022665519862

Lee, C., & Krapfl, L. (2002). Teaching as you would have them teach: An effective elementary science teacher preparation program. *Journal of Science Teacher Education, 13*(3), 247–265. https://doi.org/10.1023/A:1016521201095

Lewthwaite, B. (2014). Thinking about practical work in chemistry: teachers' considerations of selected practices for the macroscopic experience [10.1039/C3RP00122A]. *Chemistry Education Research and Practice, 15*(1), 35–46. https://doi.org/10.1039/C3RP00122A

Mahaffy, P. (2004). The future shape of chemistry education [10.1039/B4RP90026J]. *Chemistry Education Research and Practice, 5*(3), 229–245. https://doi.org/10.1039/B4RP90026J

Mahaffy, P. (2006). Moving chemistry education into 3D: A tetrahedral metaphor for understanding chemistry. Union carbide award for chemical education. *Journal of Chemical Education, 83*(1), 49. https://doi.org/10.1021/ed083p49

Marco-Bujosa, L. M., McNeill, K. L., González-Howard, M., & Loper, S. (2017). An exploration of teacher learning from an educative reform-oriented science curriculum: Case studies of teacher curriculum use. *Journal of Research in Science Teaching, 54*(2), 141–168. https://doi.org/10.1002/tea.21340

McComas, W. (2005). Laboratory instruction in the service of science teaching and learning. *The Science Teacher, 72*(7), 24–29. https://search.ebscohost.com/login.aspx?direct=true&db=eue&AN=507827291&site=ehost-live

Millar, R. (2010). Practical work. In J. Osborne & J. Dillon (Eds.), *Good practice in science teaching: What research has to say* (Vol. 2). Open University Press.

Millar, R., & Abrahams, I. (2009). Practical work: Making it more effective. *School Science Review, 91*(334), 59–64.

Millar, R., Tiberghien, A., & Le Maréchal, J.-F. (2002). Varieties of labwork: A way of profiling labwork tasks. In D. Psillos & H. Niedderer (Eds.), *Teaching and learning in the science laboratory* (pp. 9–20). Springer Netherlands. https://doi.org/10.1007/0-306-48196-0_3

Nakhleh, M. B., Polles, J., & Malina, E. (2003). Learning chemistry in a laboratory environment. In J. K. Gilbert, O. De Jong, R. Justi, D. F. Treagust, & J. H. Van Driel (Eds.), *Chemical education: Towards research-based practice* (pp. 69–94). Springer Netherlands. https://doi.org/10.1007/0-306-47977-X_4

Norris, S. P., & Phillips, L. M. (2003). How literacy in its fundamental sense is central to scientific literacy. *Science Education, 87*(2), 224–240. https://doi.org/10.1002/sce.10066

Olufsen, M., Stojanovska, M., & Petrusevski, V. (2015). Practical work in chemistry, its goals and effects. In I. Maciejowska & B. Byers (Eds.), *A guidebook of good practice for the pre-service training of chemistry teachers* (pp. 87–106). Faculty of Chemistry.

Online-Resources-for-Chemical-Safety-in-Science-Education. (2022, December 5). *Labelling*. https://chesse.org/labelling-storage-waste-management/labelling/

Online-Resources-for-Chemical-Safety-in-Science-Education. (2023, January 3). *Waste management*. https://chesse.org/labelling-storage-waste-management/waste-management/

Online-Resources-for-Chemical-Safety-in-Science-Education. (n.d). Retrieved 07.02.2023 from https://chesse.org/about-chesse/

Osborne, J. (2015). Practical work in science: Misunderstood and badly used? *School Science Review, 96*(357), 16–24.

Ottander, C., & Grelsson, G. (2006). Laboratory work: The teachers' perspective. *Journal of Biological Education, 40*(3), 113–118. https://doi.org/10.1080/00219266.2006.9656027

Pearson, P. D., Moje, E., & Greenleaf, C. (2010). Literacy and science: Each in the service of the other. *Science, 328*(5977), 459–463.

Pereira, S., Rodrigues, M. J., & Vieira, R. M. (2020). Scientific literacy in the early years—Practical work as a teaching and learning strategy. *Early Child Development and Care, 190*(1), 64–78. https://doi.org/10.1080/03004430.2019.1653553

Piaget, J. (1980). The psychogenesis of knowledge and its epistemological significance. In M. Piattelli-Palmarini (Ed.), *Language and learning* (pp. 23–24). Harvard University Press.

Sadeh, I., & Zion, M. (2012). Which type of inquiry project do high school biology students prefer: Open or guided? *Research in Science Education, 42*(5), 831–848. https://doi.org/10.1007/s11165-011-9222-9

Schleicher, A. (2016). What knowledge, skills and character qualities do successful teachers require? In A. Schleicher (Ed.), *Teaching excellence through professional learning and policy reform: Lessons from around the world*. OECD Publishing. https://doi.org/10.1787/9789264252059-3-en

Sjöström, J. (2011). Towards Bildung-oriented chemistry education. *Science & Education, 22*(7), 1873–1890. https://doi.org/10.1007/s11191-011-9401-0

Toplis, R. (2012). Students' views about secondary school science lessons: The role of practical work. *Research in Science Education, 42*(3), 531–549. https://doi.org/10.1007/s11165-011-9209-6

Utdanningsdirektoratet. (2020a). *Læreplan i kjemi (KJE01-02)*. Utdanningsdirektoratet. Retrieved 02.12.2021 from https://www.udir.no/lk20/kje01-02

Utdanningsdirektoratet. (2020b). *Læreplan i naturfag*. https://www.udir.no/lk20/NAT01-04

van Driel, J. H., Beijaard, D., & Verloop, N. (2001). Professional development and reform in science education: The role of teachers' practical knowledge. *Journal of Research in Science Teaching, 38*(2), 137–158. https://doi.org/10.1002/1098-2736(200102)38:2<137::AID-TEA1001>3.0.CO;2-U

van Driel, J. H., Berry, A., & Meirink, J. (2014). Research on science teacher knowledge. In N. G. Lederman & S. K. Abell (Eds.), *Handbook of research on science education* (Vol. II, pp. 862–884). Routledge.

Vygotsky, L. S. (1962). *Thought and language*. MIT Press.

Wang, T., Wang, W., & Wei, J. (2022). Challenges encountered by student teachers in an inquiry-based laboratory process. *Journal of Chemical Education, 99*(12), 3954–3963. https://doi.org/10.1021/acs.jchemed.2c00334

Wei, B., Chen, S., & Chen, B. (2019). An investigation of sources of science teachers' practical knowledge of teaching with practical work. *International Journal of Science and Mathematics Education, 17*(4), 723–738. https://doi.org/10.1007/s10763-018-9886-y

Wei, B., & Liu, H. (2018). An experienced chemistry teacher's practical knowledge of teaching with practical work: the PCK perspective [10.1039/C7RP00254H]. *Chemistry Education Research and Practice, 19*(2), 452–462. https://doi.org/10.1039/C7RP00254H

Wellington, J. (2002). Practical work in science: Time for a reappraisal. In J. Wellington (Ed.), *Practical work in school science* (pp. 17–29). Routledge.

Wellington, J., & Ireson, G. (2013). *Science learning, science teaching* (3rd ed.). Routledge.

Teaching Chemistry with Technology 8

Abstract

The development and progress of human society hinges on our ability to continuously develop knowledge, skills and different technologies with which we explore and effectively use natural resources. Technology has and continues to transform every aspect of our lives. Our economies are technologically driven as are our social and health systems. Education as a driver of development and societal participation has been equally transformed by technology. This chapter is about teaching and learning chemistry with available and applicable technologies. Available technologies with potential use in chemistry classrooms are explored and discussed, as well as how chemistry teachers can exploit such technologies, integrate and use them in teaching. The chapter also presents feasible actions teachers can take to develop their knowledge for teaching with technology (namely, the Teacher's Technological Pedagogical Content Knowledge).

8.1 Integrating Technology in Chemistry Teaching

The development and progress of human society hinges on our ability to continuously develop knowledge, skills and different technologies with which we explore and effectively use natural resources. Technology has and continues to transform every aspect of our lives. Our economies are technologically driven as are our social and health systems. Education as a driver of development and societal participation has been equally transformed by technology. Integration of technology in education has been long ongoing and could be traced several decades back, even though there are no documented formal discussions regarding the use of technology in teaching before the early twentieth century. Moreover, such discussions have also evolved over time along with technological advancements. For example,

F. Kayima, *Preparing for Chemistry Teaching*, Springer Texts in Education,
https://doi.org/10.1007/978-3-031-97122-8_8

from the use of educational films and radios in the early 1900s, television and computers around the 1950s, the introduction of educational software between 1950 and 1970s, and personal computers in classrooms around the 1980s, internet and e-learning in the early 1990s, the use of mobile devices, development and use of learning management systems in late 1990s (early 2000s), and then later around the 2010s and afterwards, the application of Artificial intelligence and development of advanced learning technologies, programming and gaming software, powerful language models such ChatGPT, Co-pilot, and other digital assistants and social networks.

The central role and necessity of technology in education even became more pronounced when the world faced the COVID-19 pandemic. Remote and blended or hybrid learning seemed the only viable option at the time and several developed digital communication platforms and applications such as Zoom, Microsoft teams, google meet and several social network systems (Facebook, X (formerly twitter), etc.) were central to keeping societies connected and businesses running. This chapter is about teaching and learning chemistry with available and applicable technologies. I consider and discuss the available technologies with potential use in chemistry classrooms and how teachers can exploit these and integrate them in teaching. Moreover, the chapter also discusses some ways how the teachers' knowledge for teaching with technologies coined as the teacher's technological pedagogical content knowledge (TPCK), could be developed.

8.1.1 Available Technologies Applicable in Chemistry Teaching

Interest in exploring relevant technologies for use and those already being applied in teaching has increased in recent years. In chemistry education for example, we now have several research reports (empirical, theoretical and review reports) providing an overview of the different forms of technologies relevant for chemistry teaching and learning. Examples of these include a report by Aroch et al. (2024) where they explore different modes of technology integration in chemistry, a report by Gulińska (2009) about using new technologies in teaching chemistry, a report by Ryan (2019) exploring the integration of technology in the chemistry classroom and laboratory, a report by Tuvi-Arad and Blonder (2019) focusing on teaching chemistry with databases, a book by Gupta-Bhowon et al. (2009) exploring different aspects technology application in chemistry education, a report by Mishra and Koehler (2006) advancing a framework for the teachers' technological pedagogical content knowledge (TPCK), and so many other reports. In other words, we can confidently pride ourselves on the steadily growing body of knowledge on teaching with technology in education in general, and chemistry education in particular.

A simple classification of technologies relevant in chemistry is provided by among others Tuvi-Arad and Blonder (2019) and Aroch et al. (2024). In both cases, the authors distinguish between generic modes of using technology (technologies for general educational uses) and chemistry specific modes of using technology (e.g., specific uses of technology in chemistry). Examples given in the first category include technologies for organizing learning materials such as learning management systems (LMS)—Moodle talent LMS, WebEX, Canvas, Blackboard, etc., technologies for establishing interactions between teacher and students and between students such as clickers, web-based communication systems, Microsoft teams, zoom platform, social network sites (SNS), like Facebook, google video/ chat, etc. Examples in the latter category include technologies that support visualization of submicroscopic structures, digital chemistry databases and computer simulations and computation chemistry software and apps/websites. In Table 8.1, I have included more examples of available technologies under each one of the two general categories. Moreover, as shown in Table 8.1, a categorization based on purpose, function, and area of chemistry use or focus, yields more categories of technologies available but also relevant to chemistry teaching and learning. As you will notice in Table 8.1, one technological tool can serve more than one purpose or function and hence appears under several categories. Among other included technologies are those concerned with laboratory management, chemical safety and management, technologies with a direct contribution to ensuring a safer interaction between chemistry, students, teachers and the environment, that is, technologies that deal with health, safety and environment when working with chemicals. AI -based language models and AI -powered assistant tools are also included in the table. These are technologies offering a wide range of uses in so many areas including in chemistry education.

Note that not all technologies included in Table 8.1 may be applicable to chemistry teaching situations and more so to lower levels of chemistry teaching and learning. Some technological tools serve specialized purposes and may have very little to no relevance in the day-to-day chemistry teaching at both higher and lower secondary classes, while other technologies require advanced skills and mastery to be used, which skills most teachers don't or might not have.

For example, synthesis planning tools such as SynRoute and Synthia may not be of good use at lower levels of chemistry learning. SynRoute is for example a retrosynthetic planning software tool. It uses available general reaction templates and searches available digital reaction databases to find practical synthetic routes for compounds of interest (Latendresse et al., 2023). Such operations are not included in chemistry curricula at lower levels and are hardly included in high school chemistry curricula. Thus, a chemistry teacher at either of these two levels might not have strong interest in using this retrosynthetic planning tool, however, the tool might be (and indeed) relevant at university level where further advances in understanding chemistry are pursued. Similarly, teachers at lower levels of chemistry teaching might find less use for technologies in the function category of for example, Spectroscopy and Analytical Tools¸ Artificial Intelligence and Machine Learning, and Bioinformatics and Computational Biology. This does not mean that you as a chemistry teacher, should not seek to learn about these technologies. You might be surprised on the contrary when you find out that you could indeed use some technologies under these categories in your teaching.

Table 8.1 Modes of using technology in chemistry

Modes of using technology in chemistry	Based on purpose and function and area of focus	Specific examples
General technological tools	Learning management systems	Moodle, canvas, Blackboard, google classroom, TalentLMS, SAP Litmos, Edmodo, Schoology, etc
	Interactive communication platforms	Microsoft teams, google meet, Zoom, Padlet, Kahoot, Flip, Remind, etc., social network systems such as facebook, LinkedIn, Tiktok, Snapchat, X (formally Twitter), Whatsapp, etc.,
	AI -based language models and AI -powered assistant tools	ChatGPT (Open AI), google BERT, Albert, microsoft copilot, DeepSeek, etc
	Search engines	Google, google scholar, Bing, DuckDuckGo, Yahoo! Search, Yandex, Ask.com, etc
Technological tools of specific importance in chemistry	Digital chemical databases	PubChem, ChemSpider, Reaxys, The Cambridge Structural Database (CSD), ChemEd DL (Chemical Education Digital Library), MERLOT Chemistry, RSC ChemSpider SyntheticPages, SDBS (Spectral Database for Organic Compounds), NIST Chemistry WebBook, Wolfram Alpha (Chemical Data), WebElements, SciFinder, Chemical Abstracts Service (CAS) Search, etc
	Simulation and Virtual Laboratories	PhET Interactive Simulations, Labster, ChemCollective Virtual Labs, Molecular Workbench, Beyond Labz, ChemReaX, SimPop Molecular, Virtual Chemistry Laboratory, Vernier Graphical Analysis, ChemDoodle, etc

(continued)

Table 8.1 (continued)

Modes of using technology in chemistry	Based on purpose and function and area of focus	Specific examples
	Computational chemistry and molecular modeling	Gaussian, Schrödinger Suite, ORCA, MOPAC, VMD (Visual Molecular Dynamics), GAMESS (General Atomic and Molecular Electronic Structure System), Q-Chem, Avogadro, NWChem, Spartan, MOLCAS, HyperChem, Psi4, WebMO, Tinker, MOPAC, GAMESS etc
	Spectroscopy and Analytical Tools	NMRShiftDB, SpectraBase: PyMOL, VMD, ChemDoodle, MestreNova:
	Data Analysis and Cheminformatics	KNIME, Pipeline Pilot¸ RDKit, Open Babel
	Chemical Synthesis and Reaction Design	SynRoute, Synthia, Reaxys, SciFinder, AutoSyn, ASKCOS,
	Laboratory Management and Automation	LabWare LIMS, ELN (Electronic Lab Notebooks), Chemspeed, Hamilton Robotics
	Chemical Safety and Regulatory Compliance	Chemwatch, SDS Online, REACH Compliance Software, Prism
	3D Printing and Materials Chemistry	Tinkercad, Blender, 3D Bioprinting Solutions, OpenSCAD
	Artificial Intelligence and Machine Learning	IBM RXN for Chemistry, DeepChem, AlphaFold
	Environmental Monitoring and Green Chemistry	Air Quality Monitoring Software, Green Chemistry Metrics Software, Environchem
	Bioinformatics and Computational Biology	Rosetta, AutoDock, Biovia Discovery Studio, GROMACS
	Visualization digital tools	ChemDraw, Avogadro, Jmol, PyMOL, Chem3D, VMD (Visual Molecular Dynamics), BioVia Discovery Studio, Gaussian View, Spartan, Maestro, ChemSketch, Tinker¸ ChemTools, PubChem Sketcher, MolView, WebMO, Cheminformatics Tools (ChemTools), NCI/CADD Chemical Identifier Resolver, BioDigital Human, Marvin JS,

Let us take for example the category, "spectroscopy and analytical tools" shown in Table 8.1. This is most suitable for university chemistry teaching and learning, where analysis of chemical compounds and materials by employing available instrumentation (spectral data and analytical methods) is most pronounced. However, it is possible for chemistry teachers at lower education levels (lower and upper secondary) to use technologies under this category to serve suitable teaching and learning needs. NMRShiftDB is an NMR database for organic structures and their nuclear magnetic resonance (nmr) spectra for example. However, in addition to allowing spectrum prediction and searching for spectra, the tool also provides information on structures of compounds and properties of interest. It is thus possible for a chemistry teacher in secondary school to use the tool to explore this latter interest which best aligns with chemistry curricula goals at the secondary level of chemistry. PyMOL is another example. PyMol is a molecular visualization system developed by Schrödinger Inc., that can produce high-quality 3D images of small molecules and biological macromolecules (e.g., proteins). Teachers can use this tool to help learners visualize complex structures, engage with 3D models for specific molecules interactively to explore structural formation and orientation of these structures, among other potential applications. In other words, most of the listed technological tools can be useful in chemistry teaching and learning dependent on the teacher's knowledge about the availability of the technology as well as his/her competence (both technical and didactical) as needed to employ the technology.

To bring these technologies closer to chemistry teaching contexts, I will in the next section select a few of these technological tools listed in Table 8.1 and discuss how the concrete contributions of these respective tools in chemistry teaching look like. It should be noted however that while a number of these technological tools are freely available for use without having to purchase a user license, there are those for which access is purchased. Table 8.2 is a summary of the technologies that I found to be freely accessible. Take note however that access to these tools might change over time from being freely accessible to requiring payment before use or access.

8.2 Some Concrete Examples

8.2.1 Use of PhET Simulations in Chemistry Teaching and Learning

The PhET Interactive Simulations project founded in 2002 by Nobel Laureate Carl Wieman at the University of Colorado Boulder has developed several interactive simulations for teaching and learning in several subjects, physics, chemistry, math, earth science and biology. There are over 30 interactive simulations for chemistry teaching and learning. Moreover, the PhET resource package also includes a PhET teaching ideas page at https://phet.colorado.edu/en/teaching-resources, with several activities—lesson plans and activities showing how to integrate and use PhET

Table 8.2 Technological tools with free access

Modes of using technology in chemistry	Based on purpose and function and area of focus	Technological tools with free or partial free access
General technological tools	Learning management systems	Moodle
		Google classroom
		Canvas LMS—free to use
		Schoology—free version
	Interactive communication platforms	Microsoft teams—free version
		Google meet
		Zoom—free version
		Padlet—free version
		Kahoot—free version
		Flip
		Remind
		Most SNS (facebook, tiktok, etc.) are free
	AI -based language models and AI -powered assistant tools	OpenAI's ChatGPT—free version; DeepSeek AI; etc
		Gemini
	Search engines	All generally freely accessible
Technological tools of specific importance in chemistry	Digital chemical databases	PubChem
		ChemSpider
		MERLOT Chemistry
		RSC ChemSpider SyntheticPages
		SDBS (Spectral Database for Organic Compounds)
		NIST Chemistry WebBook
		WebElements
		Wolfram Alpha (Chemical Data)
	Simulation and Virtual Laboratories	PhET Interactive Simulations
		ChemCollective Virtual Labs
		Molecular Workbench
		ChemReaX
		Vernier Graphical Analysis (Free Version)
		SimPop Molecular

(continued)

Table 8.2 (continued)

Modes of using technology in chemistry	Based on purpose and function and area of focus	Technological tools with free or partial free access
	Computational chemistry and molecular modeling	ORCA
		VMD (Visual Molecular Dynamics)
		GAMESS (General Atomic and Molecular Electronic Structure System)
		Avogadro
		NWChem
		Psi4
		WebMO (basic)
		Tinker
	Spectroscopy and Analytical Tools	NMRShiftDB
		SpectraBase
		VMD (Visual Molecular Dynamics)
	Data Analysis and Cheminformatics	RDKit
		KNIME (free version)
		Open Babel
	Chemical Synthesis and Reaction Design	ASKCOS
	Laboratory Management and Automation	Some Electronic Lab Notebooks (ELNs) are free but with limited features
	Chemical Safety and Regulatory Compliance	SDS Online
	3D Printing and Materials Chemistry	Tinkercad
		Blender
		OpenSCAD
	Artificial Intelligence and Machine Learning	IBM RXN for Chemistry
		DeepChem
		AlphaFold
	Environmental Monitoring and Green Chemistry	
	Bioinformatics and Computational Biology	AutoDock
		GROMACS
	Visualization digital tools	Avogadro
		Jmol

(continued)

Table 8.2 (continued)

Modes of using technology in chemistry	Based on purpose and function and area of focus	Technological tools with free or partial free access
		PyMOL
		VMD (Visual Molecular Dynamics
		ChemSketch
		MolView:
		PubChem Sketcher
		Marvin JS
		WebMO
		NCI/CADD Chemical Identifier Resolver

simulations in teaching. In addition to providing access to different teaching activities that integrate and use PhET simulations, this teaching resources page also provides Tips for using PhET simulations and allows different teachers and educators worldwide to upload and share their teaching activities that integrate and use PhET simulations. The page also provides for activities in different languages to be uploaded and shared and thereby provides the opportunity for all interest groups regardless of their geographical location to exploit and use the shared teaching ideas.

The PhET project team developing these simulations aims to support students to engage in scientific exploration, take ownership of learning experiences, develop conceptual understanding, connect concepts and learning to everyday life, and to view science as accessible and enjoyable (Moore et al., 2014). Depending on the teacher's teaching goals, interest and competence, PhET simulations can be used in several ways to suit one's teaching needs. Moore and colleagues (2014), discuss at least two ways of integrating and using PhET simulations in chemistry teaching. The first is when the teacher uses the simulations as complementary learning aids in a traditional teacher-led classroom session. That is, by for example projecting a selected simulation onto a screen and demonstrating the different aspects of interest for students to visualize the sought relationships.

This way of using PhET simulations is framed within the teaching itself, and the simulations are used to illustrate the ideas being addressed in the main teaching. The teacher can give students the opportunity to play with or manipulate PhET simulations by providing guiding tasks to work with during the lesson. For instance, a teacher, having introduced the concept of isotopes and abundance, can use PhET simulations to demonstrate the relationship between number of electrons, number of protons, number of electrons, mass number, atomic number and the atomic mass (amu). By selecting an element from the provided periodic table, a student gets to see what isotope is most stable, its corresponding abundance in nature, what constitutes mass number and the distinction between mass number

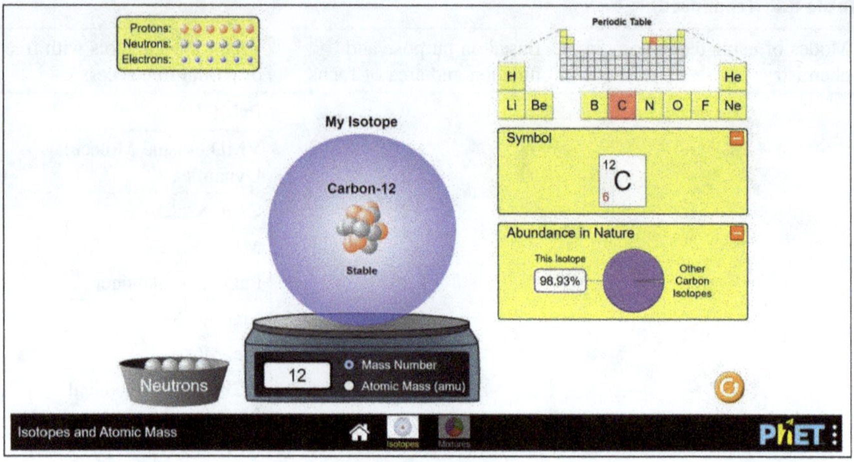

Fig. 8.1 Screenshot from https://phet.colorado.edu/sims/html/isotopes-and-atomic-mass/latest/isotopes-and-atomic-mass_all.html; Simulation by PhET Interactive Simulations, University of Colorado Boulder, licensed under CC-BY-4.0 (https://phet.colorado.edu)

and atomic mass (amu). S/he can also explore the possible isotopes for a given element by adding and or deducting neutrons from the "my Isotope" and then reading off the provided abundance in percentage (see Fig. 8.1).

The second approach is the guided inquiry approach. With this approach, the teacher provides information that guides the students as they explore selected PhET simulations. Students can work in groups where each group member has a defined role to ensure that all students are involved actively. To make the inquiry meaningful, students need to have some background knowledge about the topic so that they do not start from scratch. For example, the PhET interactive simulations assume that students are already familiar with the language and terms that are used in these simulations. Students should be able to relate to the chemical language and terminologies used in the different simulations or otherwise have the expectation that they will have to explore and find appropriate meanings to the used terminologies in the explored simulations. Different tasks can be given to students to guide them in the inquiry activity. These could for example range from open-ended tasks that ask students to explore, manipulate selected simulations, make sense of visualized relationships, and develop assertions to reflect their understandings that have emerged, or they could be more concrete tasks requiring students to make conclusions about certain visualized relationships. Blow (Table 8.3) is an example of an activity that gives students the freedom to explore and manipulate a selected simulation but with concrete guiding tasks that will lead students towards specific target learning goals. Moreover, as already mentioned, there are several developed teaching activities accompanying each one of the 30 developed PhET interactive simulations for chemistry. For the "Balancing Chemical Equations" for example,

Table 8.3 Guided inquiry activity: Understanding the law of mass conservation through balancing chemical equations

Activity	
Learning goals	Explore and understand the law of conservation of mass through balancing chemical reactions
	Distinguish between coefficients and subscripts and their respective functions in molecules and compounds taking part in chemical reactions
Guiding questions	Use a few minutes to explore the Balancing Chemical Equations screen. Find out what different controls do or show
	Select the option "Make Ammonia" and then work towards balancing the showed chemical equation
	Repeat the process for the options "Separate Water" and "Combust Methane" and in each make notes about the different coefficients needed both for the side of reactants and products
	Select the "Game" tab/option and select a level of your choice. Use a few minutes to play the game of balancing the provided chemical questions. If you have chosen to start with level 1, try to quickly work through until you reach and solve tasks at level 3
	Based on the explorations you have done, discuss the following questions 1. What roles do both coefficients and subscripts serve in a well-balanced chemical equation? 2. Discuss what a well-balanced chemical equation can inform about the law of conservation of mass

these are accessed at https://phet.colorado.edu/en/simulations/balancing-chemical-equations/activities.

8.2.2 Using ChemCollective Resources to Teach and Learn Chemistry

The ChemCollective (https://chemcollective.org/home) is another digital library of online activities for teaching and learning chemistry. This rich digital library contains a variety of ready to use activities including, Virtual lab problems that use a flexible simulation of a wet chemistry lab, auto-graded Virtual lab problems that embed virtual lab into a webpage that allows randomized generation of unknowns and automatically checking of student's answers, tutorials combining movies and text with interactive problems, scenario-based learning activities that give students the opportunity to apply chemistry concepts in a real-world situation, particulate level visualizations connecting chemistry central ideas to molecular motions and interactions, ready to use concept tests for assessment for learning purposes, and fully fledged online courses offering comprehensive instruction through videos, text, virtual lab activities, practice tutors and assessments (https://chemcollective.org/teachers/index). Each one of these mentioned resources is freely accessible, licensed under a Creative Commons Attribution 3.0 NonCommercial-NoDerivs

License as of the 19th of September 2024, and it is most likely that free access to these resources will continue if the team developing these resources has the means to allow free access.

Of the above-mentioned ChemCollective resources, we take a closer look at the Virtual lab as one example. The included virtual lab activities cover a range of experiments for both high school and college/university general chemistry. According to the developers, the Virtual lab is designed to allow students perform calculations as they would in lab or use the pH meter/concentration tables to check their understanding of the system. It also allows students to design and carry out their own experiments (Yaron et al., 2010). Figure 8.2 is a screenshot of the Chemcollective Virtual lab (used with permission). The screen is divided into three parts, on the left is the stockroom for chemical reagents and other fictional materials with different properties aligning with the teacher and students' interest. The middle part is the work area known as the workbench. This is where experiments are performed. Note that in the working area (workbench), there are four icons on the top left that serve different purposes. For example, one icon provides access to lab equipment such as beakers, conical flasks, measuring cylinders, pipettes, etc., while the other icon provides access to equipment such as the Bunsen burner, weighing scale/boat, etc. On the right of the screen, you find a multiple representation of the contents selected to be used in the workbench area. This includes information about the pH of the selected solutions, temperature, amount in moles, grams, molar concentration, etc. (Yaron et al., 2010).

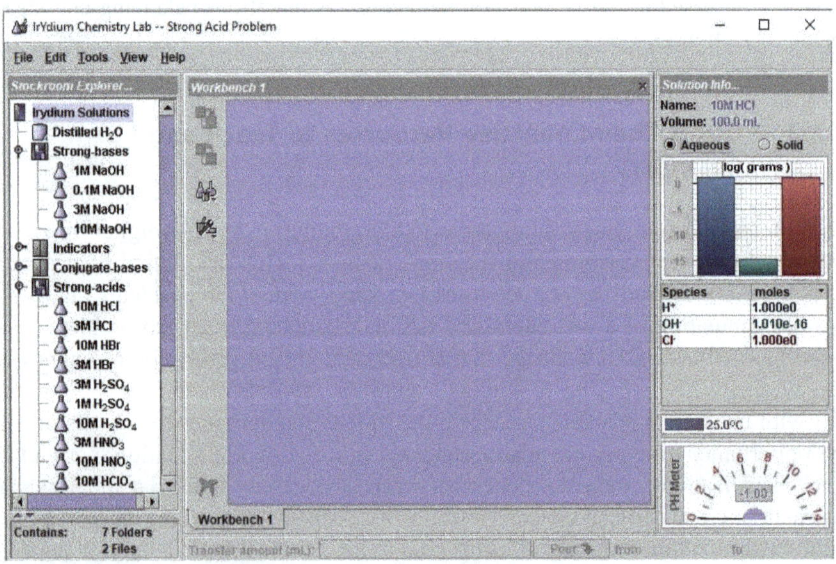

Fig. 8.2 Screeshot of the ChemCollective Virtual lab—used with permission

As the developers of these ChemCollective resources argue, the virtual lab component offers laboratory experiences like those performed in a physical laboratory. This provides opportunities for all students to conduct different experiments virtually, going through the experimental processes as if they were in a physical laboratory. Moreover, it allows students to perform all necessary calculations that go hand in hand with completion of target experiments. Moreover, they offer a collection of pre-written problems that students can work with either as integrated classroom activities or as home tasks and the virtual lab can be used either online or offline by downloading it to a suitable and compatible gadget https://chemcollective.org/vlab_download.

Like for the PhET simulations, to use ChemCollective resources in your chemistry teaching, the teacher needs to take time to explore and learn how these different resources are used, how they do work, what kind of tasks are provided and how these are solved, etc. S/he needs to consider these resources in terms of what is relevant in the chemistry curriculum that guides what is to be taught about chemistry in his/her context. Most importantly, the teacher needs to learn to manipulate these digital tools, navigate through the different resources since he/she will need to help or support his/her students once they start to use the same resources. In Table 8.2, under the subcategory simulation and virtual laboratories, other resources offering similar chemistry learning opportunities and experiences are listed, including the ***Molecular Workbench*** which provides visual and interactive computational experiments (http://mw.concord.org/modeler/) and ***the ChemReax*** web app by ScienceBySimulation for modelling and simulation of chemical reactions. The ChemReaX app also offers experimental experiences in acid–base titrations and includes a list of virtual lab exercises focusing on selected chemistry topics (https://www.sciencebysimulation.com/chemreax/Analyzer.aspx). All these resources can be accessed by visiting the respective websites.

8.2.3 Using Digital Chemical Databases in Chemistry Classrooms

The amount of chemical content is vast and steadily growing and the rising number of developed digital chemical databases has been a successful response to the challenges of managing such massive volume of chemical information that we have today. Each one of the existing digital chemical databases as shown in Tables 8.1 and 8.2, addresses or focuses on specific areas of chemistry. The PubChem (https://pubchem.ncbi.nlm.nih.gov/) for example, is a public repository of information with a focus on small molecules and their biological activities, while ChemSpider (https://legacy.chemspider.com/) is a chemical structure database for over 129 million chemical structures from about 270 data sources. Digital chemical databases serve a number of functions, including facilitating and allowing an easy and smooth handling of large volumes of chemical data—e.g.; millions of compounds, reactions and properties, allowing timely retrieval of this data through integrated technology capabilities, providing for organizing, verifying, and updating chemical data to provide an accurate and reliable storage source of chemical information, and thereby facilitating research and further development of chemical knowledge, among other purposes. Moreover, digital chemical databases are resources that support education and learning in chemical sciences. Indeed, our greatest interest as teachers lies in the potential teaching and learning benefits offered by these existing digital chemical digital databases. There are a range of ways chemistry teachers can exploit existing digital chemical databases to enhance chemistry learning experiences. From the freely accessible digital databases in Table 8.2, let us consider and briefly discuss three of these databases, namely, the WebElements, PubChem, and ChemSpider, regarding how each one of these could be used in chemistry teaching contexts to enhance students' chemistry learning experiences.

8.2.3.1 The WebElements™

The WebElements™ (https://www.webelements.com/) created by Mark Winter in 1993 provides detailed information about chemical elements and their properties. It provides comprehensive data about each member of the periodic table, including the chemical element's atomic properties, physical and chemical properties, the historical information of the element, as well as any other known information about a given element such its biological and environmental role, industrial and everyday applications. Many science educators who have explored the WebElements™ digital resource and compared it with other similar resources that cover chemical elements, and their properties conclude that information provided by WebElements™ is not only comprehensive but has a higher degree of accuracy, and easy to access and use (Tuvi-Arad & Blonder, 2019). Moreover, the WebElements™ online resource also provides some information about compounds formed by elements and the corresponding reactions, and it offers visual and graphical information to ease the exploration of trends in the properties of chemical elements.

With its several features, teachers can use the WebElements™ digital resource to introduce students to the periodic table by creating activities that give students the opportunity to explore the chemical properties of given elements and the periodic trends. Moreover, teachers can plan guided inquiry tasks targeting specific properties and periodic trends where students explore and navigate the periodic table and through the information, including the graphic presentations as they work towards solving the provided tasks. The interactive experience with the digital periodic table thus offers an engaging approach to learning periodicity. A teacher could for example task students to explore binary compounds of chemical elements using the digital WebElements™ periodic table. The activity can be designed in such a way that students work in small groups, where each group chooses one element within, for example group 1 and 2 and 13 and then proceed to explore the different possible compounds for the selected chemical element. Since for each selected chemical element, there are several compounds starting for example from Lithium hydrides, fluorides, chlorides, bromides, iodides, oxides, etc., it is important for the teacher to limit the scope of the exploration and to provide concrete guidance on what to exactly focus on, so that students are not overwhelmed by the massive information available to them via this platform.

8.2.3.2 ChemSpider and PubChem

ChemSpider is a free chemical database developed by the Royal Society of Chemistry. It is free to use and does not require any signup. It is a search engine containing over 129 million structures. ChemSpider contains a myriad of different kinds of information such as physical and chemical properties of molecules and interactive spectral data. It also includes links to chemical Venda catalogues, literature references and patents, among other things. This information is collected from hundreds of different sources including Royal society of chemistry journals and databases, other publishers and indexing services, specialist databases and several internet sources such as Wikipedia (https://legacy.chemspider.com/DataSources.aspx). The database can be searched by either using the name of the compound, by drawing the structure (legacy structure search) or by using a combination of different types of searches. The large pool of chemical information provided by ChemSpider offers a wide range of use for chemistry teachers. For example, students can explore the structure of relevant molecules, visualizing both 2D and 3D structures, navigate through different properties of a given compound, learn about its possible isomers, etc. Moreover, the teacher can use the digital resource to plan guided inquiry activities that foster and thereby enhance students' problem-solving skills and critical thinking.

Like ChemSpider, PubChem is a digital collection of freely accessible chemical information but with a focus on small molecules and their biological activities. The database was developed and continues to be managed by the National Library of Medicine (NLM) since its launch in 2004. As of 23. 09. 2024, PubChem contains over 119 million compounds, 322 million substances, 295 million bioactivities, over 42 million literature and 51 million patents. Information is collected from over 1007 million data sources (https://pubchem.ncbi.nlm.nih.gov/). The database can be searched by using the name of the compound, using the molecular formula, by drawing the structure and searching for the drawn structure, or using other identifiers. The source provides access to chemical information, physical and chemical properties, biological activities, toxicity information, spectra data, patents and literature information among other things. It thus provides teaching and learning opportunities in the same way ChemSpider does.

8.3 Generative Artificial Intelligence (GAI) in Chemistry Teaching and Learning

In the introductory editorial by Yuriev et al. (2024) for the Journal of Chemical Education's 8th 2023 Special issue, the virtual special issue (VSI) titled; *"Investigating the uses and impacts of Generative Artificial Intelligence in chemistry education"* (Yuriev et al., 2023), the authors write;

> As dawn breaks on generative AI, it heralds a new era in chemistry education, pushing the traditional boundaries and opening up new pathways for pedagogical innovation. This ground-breaking technology is capable of creating content, solving problems, and engaging in sophisticated dialogue. Akin to a digital alchemist, it has the potential to "transmute" traditional teaching methodologies into more interactive experiences and revolutionize how chemistry is taught and learned. (p. 2957)

As expressed above, Yuriev et al. (2024) and many other chemical educators and scholars underscore the profound impact generative intelligence technologies have on the general educational landscape, and on chemistry and chemistry education in particular. The potential benefits and contributions by GAI technologies with regards to students' learning and the application in teaching are highlighted in several research reports (see for example; Alasadi & Baiz, 2023; Deng & Flynn, 2023; Emenike & Emenike, 2023; Exintaris et al., 2023; Guo & Lee, 2023; Hasrod et al., 2024; Mahapatra, 2024). Some of these include GAIs helping to create personalized learning experiences, providing real-time feedback to students' queries, being used by instructors to create assessment items, scoring rubrics, and interactive learning activities, in preparation of teaching materials, supporting in-class problem solving, having the potential to provide support for students with needs such as overcoming language barriers, and providing research related and academic writing support. Moreover, several educators have argued that working with GAI Chabot systems like ChatGPT can enhance students critical thinking skills, help students develop and improve their programming skills, and improve their conceptual understanding of scientific concepts.

Considering the above claimed potential benefits and contributions of GAI technologies to the educational landscape, I made a brief review of the 39 articles published by the journal of chemical education, the articles which comprise the Journal's special issue,—the Virtual Special Issue (VSI) titled, *investigating the uses and impacts of Generative Artificial Intelligence in chemistry education"*, (Yuriev et al., 2023). The VSI aimed to highlight the emerging applications of GAI in chemistry education and according to the Issue's editors (Yuriev et al., 2024), the set of the 39 articles presented in the VSI revealed a robust range of options for using GAI technologies in chemistry education as well as the potential for misuse of these technologies, and hence a guide in the use of GAI technologies in chemistry education going forward. My brief review of the 39 articles comprising the VSI, focused on identifying the different possibilities, or opportunities presented in these articles to chemistry teachers who use or intend to use GAI technologies in chemistry teaching, and the inherent challenges and limitations of current GAI systems. Different study aims were covered in the respective 39 different studies that comprise the VSI, including exploring the potential impacts of GAI text- based systems for the chemistry community (Emenike & Emenike, 2023), exploration of the role of GAI in education (Alasadi & Baiz, 2023), ChatGPT's ability to generate laboratory reports and essay and scientific writing (Humphry & Fuller, 2023; Reddy et al., 2024; Rojas, 2024; Trčková et al., 2024; West et al., 2023), evaluation of ChatGPT's generated responses to exams and other kinds of chemistry questions and for generating and scoring exams (Clark, 2023; Clark et al., 2023;

Fergus et al., 2023; Fernández et al., 2024; Leon & Vidhani, 2023; Ruff et al., 2024; Sorenson & Hanson, 2024; Tyson, 2023), use of GAIs in lesson planning and teaching (Clark et al., 2024; Gonçalves Costa et al., 2024; Jones et al., 2024; Lear, 2024; Yik & Dood, 2024), use of GAIs to foster students' critical thinking (Exintaris et al., 2023; Guo & Lee, 2023; Pence et al., 2024), students perceptions of GAIs and their use in teaching and learning contexts (Mendez, 2024; Young et al., 2024), the ability of GAI to perform chemistry calculations correctly (Clark et al., 2023; Schrier, 2024), etc.

Conclusions from most of the above studies revolve around what the explored GAI was able to do or execute successfully and that which it was not able to do or execute with 100% success, and what this means in terms of chemistry teaching and learning. For example, Emenike and Emenike (2023) concluded that AI-generated text can be used by both teachers and students for mentoring, research and professional activities, and that while it gives factual chemical information that tends to be correct, answers to specific questions come with errors and misleading submissions. West et al. (2023) concluded that ChatGPT version 3.5 was not capable of producing higher-level lab reports, while Tyson (2023) expressed that «ChatGPT cannot perform mathematical operations reliably, makes conceptual errors, and fabricates plausible-looking citations, the purported contents of which are only partly accurate." (p. 3098). Some others like Clark (2023); and Fergus et al. (2023) conclude that certain GAIs like ChatGPT have the capability to interpret chemical symbolism, conceptualize chemical problems, understand and use chemical language that is understandable to both teacher and students, and are capable of conducting a conversation with whoever engages with the system. On the other hand, the authors note that these systems make numerous errors, describe incorrect problem-solving strategies, omit information, produce responses with visible misconceptions, mathematical errors, and fail to perform basic chemistry tasks including solving problems that involve drawing, structural presentations, graphical and spectral data.

However, those educators who explored newer versions of GAIs, especially the Multimodal GAIs conclude with a different picture about what these newer versions can do. Schrier (2024) for example used the more advanced ChatGPT-4 model to find out if he would come to a similar conclusion as Clark (2023) who had used the single-modal ChatGPT-3.5 free version. Schrier concludes that indeed GPT-4 outperforms both GPT-3.5 and human students, and this level of success is attributed to both the enhanced capabilities of the newer version of the model and the teacher's ability to use correctly formulated prompts and an external support tool such as a calculator tool.

Schrier's (2024) emperical findings underscore the potential contributions of advanced GAI systems to chemistry education. Moreover, the findings seem to counter one of the strongest criticism made so far by some educators expressing for example that students will be illserved by using GAIs in their chemistry learning endevour. An important aspect that we must note in all these studies is that those studies that used or explored earlier GAI versions, such as ChatGPT-3, or 3.5, which are indeed classified as single -modal GAIs, report numerous errors and limitations of the model in executing chemistry related problems and tasks. On other hand, advanced multimodal GAI versions have advanced capabilities including image input capabilities that allow the system to solve problems that involve pictures, diagrams or any other graphical information. These newer systems can analyze scientific images with accuracy, identify graphical data points, spectral data, and any other visual representation (Alasadi & Baiz, 2024). Thus, the potential educational benefits of GAIs become much more visible and realizable with advanced versions and the productivity of GAI systems in educational contexts will continue to increase as newer modfications are being made to existing GAI systems.

What the above background implies to us chemistry teachers, educators and trainees is that GAIs are a powerful tool that we can exploit in a variety of ways as described and evidenced in the different empirical studies. Moreover, as you have so far realized, the limitations reported of earlier versions GAI models are being eliminated in more advanced models, a process that is poised to continue with the rapid technological advancements. It means that we should focus our attention on how we can utilize the capabilities of these GAI systems in our chemistry teaching, in planning chemistry lessons, in designing effective assessments, when we want to foster critical thinking and other relevant attributes and science skills, and much more. Moreover, chemistry teachers will need to learn to use the different capabilities of these GAI systems to be able to use them effectively. For example, as noted above, Schrier (2024) used prompts and an external calculator tool to achieve the success level he achieved with the ChatGPT-4 model. His methodology required that he had a good level of programming skills, and in particular the ability to use Mathematica (Wolfram Mathematica: Modern Technical Computing) and linking it with ChatGPT-4. This is not to imply that you cannot use these GAI systems without having programming skills. Indeed, the majority of the 39 studies mentioned above used a selected GAI model of their choice without any need for programming skills but simply typed in the different questions and queries directly into the GAI search engine and got the responses they desired (that is, they fed information into the system/model and received a corresponding response/output). It is possible to do the same. However, those with programming skills will even have more potential to explore these GAI capabilities.

8.3.1 What We Should Concern Ourselves with as We Embrace and Use GAIs in Educational Contexts—Ethical Implications and Other Disruptions

To this end, we have a good overview of the opportunities provided by GAIs in education, for example including but not limited to providing personalized learning experiences and adaptive learning materials, providing real-time feedback and assessment to students, working as language support tools, writing academic texts and essays, for creating impressions, illustrations and pictures, acting as reading partners, reading and summarizing large texts, etc. (see for example Alasadi &

Baiz, 2023). These are only a few of the several potential benefits and services offered by the advancing capabilities of GAIs. However, there are also several concerns about the potential impact of these GAIs on students' learning and educational outcomes. Some of these raised concerns relate to the academic integrity and plagiarism where students employ GAI tools to complete essays and assignments and even thesis reports, erosion of students' critical thinking and skill development resulting for overreliance on GAI tools, issues of data privacy and consent where GAI tools collect and store students' or teachers' data without any clear consent or without their knowledge, and the lack of access to all students among other concerns (Alasadi & Baiz, 2023; Emenike & Emenike, 2023; Li et al., 2023; Lim et al., 2023; Ruff & Zemke, 2025).

Taking assessment as an example, these GAI tools have substantially shaken and challenged traditional boundaries with respect to teaching and Assessment practices across disciplines and educational levels, and are poised to revolutionize these practices (Rudolph et al., 2023; Yuriev et al., 2023). The most traditional assessment forms such as the use of essays and home exams (homework) no longer provide the same effect as they did before these GAI tools emerged. Moreover, teachers and educators continue to find it difficult to develop and adapt alternative assessment frameworks or forms to replace these traditional ones or altogether develop alternative forms to take advantage of the capabilities of these language models. Emenike and Emenike (2023) posed a question as whether instructors should continue to assess students based on the clarity and formatting of descriptive assignments (e.g., lab reports, research papers, literature reviews) well knowing that there are now available AI tools to help accomplish these tasks effectively. This is a question that comes to you (chemistry teacher/teacher trainee) directly and whose answer demands that teachers consider new and alternative assessment forms, frameworks and approaches that consider and recognize GAI tools as part teaching and learning tools available to both teachers and students. Several educators in different teacher training institutions and colleges have the same question regarding assessment in the face of GAI capabilities. The response or solution however, is not straightforward, but lies in embracing a paradigm shift regarding assessment for educational purposes. Alasadi and Baiz (2023) responds to the above question by Emenike and Emenike (2023) by expressing that

> ….it is not necessary to grade students solely on their writing skills, particularly when advanced AI tools can accomplish these tasks. However, this does not imply that writing assessment should be completely disregarded. It is context-dependent, in introductory chemistry courses, focusing on students' conceptual understanding might be more beneficial. In contrast, for upper-division or graduate-level courses, especially those involving original research projects or scientific writing, assessing writing skills becomes increasingly relevant. Students' ability to convey complex ideas clearly, logically, and effectively is a skill that transcends disciplines. As with past technological innovations, we must balance the benefits and challenges that come with integrating AI into our educational practices. (p. 2967)

The authors, Emenike and Emenike (2023) argue for a contextual approach when it comes to assessment and do not support abandoning writing assessment forms. The contextual approach implies that you the teacher determines what kind of assessment will serve the purpose of a given group of students and depending on the nature of learning goals. In other words, an adaptive and thoughtful integration of GAIs in our chemistry teaching and assessment practices is and will continue to be needed.

Finally, the question of what we should be concerned about due to the increasing use of GAIs in education is not only about finding ways to deal with assessment disruptions caused by GAIs, but also about dealing with other arising ethical issues and pedagogical implications. Our students must be oriented and should understand ethical issues surrounding the use of GAI tools in fulfilling their academic needs. As a teacher, this is in part your responsibility. You will be expected to set clear guidelines and expectations for your students in relation to how they can work with or use GAI tools, teach AI literacy to your students and set an example by showing how they can use GAI tools with transparency, etc. Such a responsibility demands that a teacher acquires sufficient knowledge and a good understanding of ethical issues and pedagogical implications surrounding the use of GAIs in both teaching and research related activities. Keeping up with emerging policies that guide the use of GAI at your local institution is a starting place for you to get insights into these ethical obligations.

8.4 Teacher Knowledge for Teaching with Technology

Having explored some of the available technologies with potential applications in chemistry education, we now turn attention to yet another important aspect of teaching chemistry with technology, that is, the teacher's technological pedagogical content knowledge (TPCK) as it has been termed in educational literature. When discussing the teacher's knowledge for teaching in Chapter 2, we considered three major knowledge categories, the general pedagogical knowledge, the subject-specific didactics (pedagogical content knowledge), and the chemistry content knowledge. We further broke down and distinguished between teacher knowledge elements acquired in and with practice and teacher knowledge elements acquired or developed through undertaking college or university courses or teacher training. Moreover, we distinguished chemistry didactics from general pedagogy by emphasizing that chemistry didactics entails a set of knowledge and skills about the teaching and learning of particular subject matter (chemistry in our case) that considers the particular learning demands inherent in the subject matter. We added that this knowledge is underscored and thereby informed by four important questions, (i) what is it that is to be taught (chemistry content), (ii) How is it to be taught? (iii) How do students learn and how does learning takes place? and (iv) what is that which has been learned by students? These four questions constitute and thus reflect the different kinds of knowledge elements and skills teachers need to be able to teach. The dawn of technological advancements with direct impact on the educational landscape, implies expanding the scope of a teacher's knowledge for teaching to include and account for a knowledge element that enables the use of technology in teaching.

The effort and urge to characterize what teachers need to know to be able to teach with technology and to keep up with the emerging technological demands, is a subject that the educational community has grappled with for some time. On one hand, teachers need to have knowledge of existing technologies with potential educational applications. On the other, they need to have the ability to evaluate and select among these technologies, the most suitable ones to use in given teaching contexts. This latter expectation is about the ability to consider the nature of chemistry content aspects to be taught with their inherent and unique learning demands and select a potential technology that the teacher sees as suitable and would serve to facilitate the intended learning. Mishra and Koehler (2006) proposed what they termed a TPCK model to account for a teacher's knowledge for teaching with technology—TPCK—Technological Pedagogical Content Knowledge. They characterized TPCK as a knowledge that forms the basis for good teaching with technology, that which demands

> ….an understanding of the representation of concepts using technologies; pedagogical techniques that use technologies in constructive ways to teach content; knowledge of what makes concepts difficult or easy to learn and how technology can help redress some of the problems that students face; knowledge of students' prior knowledge and theories of epistemology; and knowledge of how technologies can be used to build on existing knowledge and to develop new epistemologies or strengthen old ones. (p. 1029)

While the importance of having this knowledge as proposed and characterized by Mishra and Koehler (2006) cannot be overemphasized, the greatest challenge lies in how teachers can build, develop and or acquire this kind of knowledge. This is the question you should be concerned with here. The whole of this chapter presents an orientation into the available modes of technologies with potential relevance to chemistry teaching and learning. How then can we as chemistry teachers acquire or develop a type of knowledge (TPCK), with which we can make decisions about what technologies are relevant in different chemistry teaching contexts, for different chemistry topics and concepts, when and how to use or apply these technologies? I will approach the solution to the above question by suggesting three actions,

First, you the chemistry teacher will need and or should explore and learn about the available modes of technologies with potential application in chemistry teaching contexts. There are several ways through which you can get to know which technologies exist today that have potential use in education. Examples include, reading current published teaching materials where the topic on teaching with technology is considered and addressed, attending workshops on ICT use in education, or short training sessions, considering the commonly available social network systems, performing online self-study where you explore the internet to find the different technologies that contribute to the educational landscape, sharing and learning from colleagues who happen to have more knowledge about the available technologies, keeping up to date with technological advancements by following news and developments and how these shape the future of education. In short, as a chemistry teacher, you should take the initiative to explore and get to know about the existing technologies that could be relevant for your teaching.

Second, start by selecting a few and then make an effort to learn how they work and how you yourself can use them. Say for example that you have selected PhET simulations, you have never used this resource before in your teaching but are interested and want to integrate this resource into your teaching. Start by exploring the different capabilities of the resource to learn what the different included functions do. You do this exploration yourself or ask a colleague familiar with the system to take you through the basics. Remember that knowing how the system works or how to use the system is not the same as being able to teach with the system. Therefore, do not rush into introducing the system in your teaching simply because you have learnt how to use it or simply because you know how the system works. You will find it beneficial to take some time to get a full understanding of the different operations of the system, reading about how others have used the system and under which circumstances, etc.

Third, it is now time for you to learn how to teach with the system. Having learnt to use the system and explored different avenues (sources and examples) where the same system has been employed or used in teaching situations, start by designing a teaching where you use parts of this system or resource that you deem useful and want to integrate into your teaching. It is important to create enough space for making mistakes and you should be clear about this expectation. Whenever possible, keep your plan simple and less complex. Mishra and Koehler (2006) argued that a teacher's knowledge for teaching with technology is most effectively attained through learning by doing, the so called "learning-technology-by-design approach. This third step here is only a part of what should be done if one is to develop this kind of knowledge. As you realize, all three steps are important for the process of learning by doing to be realized. One needs to know the available and relevant technologies, learn how these technologies work and then proceed to explore how the selected technologies can be used in teaching contexts.

Moreover, one can find different literature sources with a myriad of advice and suggestions concerning what factors to consider in selecting the technologies, how to navigate exiting barriers to technology integration among others. For example, recommendations such as focusing on and starting with those technologies which have been already tried out and tested in real teaching situations can be important, especially for the first and second phases described above. Doing so exposes you both to the available and relevant technologies and to a set of various examples of how these technologies have been used in teaching situations. Regardless of whether you choose to start with already tested or untested technologies, developing a knowledge for teaching with technology will always encompass all three phases I have described. Moreover, it is important to be prepared and open for learning new technological skills and this means going over to the first and second phases again and again owing to the rapid rate of technological change. We thus cannot afford the luxury of assuming ourselves to be fully learned and skilled in using these technologies. We must embrace the challenge of being challenged as teachers and make this part of our reality going forward. We must strive to find the motivation to learn how to use emerging technologies if we want to learn how to use them in our teaching.

References

Alasadi, E. A., & Baiz, C. R. (2023). Generative AI in education and research: Opportunities, concerns, and solutions. *Journal of Chemical Education, 100*(8), 2965–2971. https://doi.org/10.1021/acs.jchemed.3c00323

Alasadi, E. A., & Baiz, C. R. (2024). Multimodal generative artificial intelligence tackles visual problems in chemistry. *Journal of Chemical Education, 101*(7), 2716–2729. https://doi.org/10.1021/acs.jchemed.4c00138

Aroch, I., Katchevich, D., & Blonder, R. (2024). Modes of technology integration in chemistry teaching: theory and practice [10.1039/D3RP00307H]. *Chemistry Education Research and Practice, 25*(3), 843–861. https://doi.org/10.1039/D3RP00307H

Clark, T. M. (2023). Investigating the use of an artificial intelligence Chatbot with general chemistry exam questions. *Journal of Chemical Education, 100*(5), 1905–1916. https://doi.org/10.1021/acs.jchemed.3c00027

Clark, T. M., Anderson, E., Dickson-Karn, N. M., Soltanirad, C., & Tafini, N. (2023). Comparing the performance of college chemistry students with ChatGPT for calculations involving acids and bases. *Journal of Chemical Education, 100*(10), 3934–3944. https://doi.org/10.1021/acs.jchemed.3c00500

Clark, T. M., Fhaner, M., Stoltzfus, M., & Queen, M. S. (2024). Using ChatGPT to support lesson planning for the historical experiments of Thomson, Millikan, and Rutherford. *Journal of Chemical Education, 101*(5), 1992–1999. https://doi.org/10.1021/acs.jchemed.4c00200

Deng, J. M., & Flynn, A. B. (2023). "I Am Working 24/7, but I can't translate that to you": The barriers, strategies, and needed supports reported by chemistry trainees from English-as-an-additional language backgrounds. *Journal of Chemical Education, 100*(4), 1523–1536. https://doi.org/10.1021/acs.jchemed.2c01063

Emenike, M. E., & Emenike, B. U. (2023). Was this title generated by ChatGPT? Considerations for artificial intelligence text-generation software programs for chemists and chemistry educators. *Journal of Chemical Education, 100*(4), 1413–1418. https://doi.org/10.1021/acs.jchemed.3c00063

Exintaris, B., Karunaratne, N., & Yuriev, E. (2023). Metacognition and critical thinking: Using ChatGPT-generated responses as prompts for critique in a problem-solving workshop (SMARTCHEMPer). *Journal of Chemical Education, 100*(8), 2972–2980. https://doi.org/10.1021/acs.jchemed.3c00481

Fergus, S., Botha, M., & Ostovar, M. (2023). Evaluating academic answers generated using ChatGPT. *Journal of Chemical Education, 100*(4), 1672–1675. https://doi.org/10.1021/acs.jchemed.3c00087

Fernández, A. A., López-Torres, M., Fernández, J. J., & Vázquez-García, D. (2024). ChatGPT as an instructor's assistant for generating and scoring exams. *Journal of Chemical Education, 101*(9), 3780–3788. https://doi.org/10.1021/acs.jchemed.4c00231

Gonçalves Costa, G., J. D. Nascimento Júnior, W., Mombelli, M. N., & Girotto Júnior, G. (2024). Revisiting a teaching sequence on the topic of electrolysis: A comparative study with the use of artificial intelligence. *Journal of Chemical Education, 101*(8), 3255–3263. https://doi.org/10.1021/acs.jchemed.4c00247

Gulińska, H. (2009). Using New technologies in teaching chemistry. In M. Gupta-Bhowon, S. Jhaumeer-Laulloo, H. Li Kam Wah, & P. Ramasami (Eds.), *Chemistry education in the ICT age* (pp. 131–144). Springer Netherlands. https://doi.org/10.1007/978-1-4020-9732-4_15

Guo, Y., & Lee, D. (2023). Leveraging ChatGPT for enhancing critical thinking skills. *Journal of Chemical Education, 100*(12), 4876–4883. https://doi.org/10.1021/acs.jchemed.3c00505

Gupta-Bhowon, M., Jhaumeer-Laulloo, S., Wah, H. L. K., & Ramasami, P. (2009). *Chemistry education in the ICT age*. Springer.

Hasrod, T., Nuapia, Y. B., & Tutu, H. (2024). ChatGPT helped me build a chemistry app, and here's how you can make one also. *Journal of Chemical Education, 101*(2), 653–660. https://doi.org/10.1021/acs.jchemed.3c01170

Humphry, T., & Fuller, A. L. (2023). Potential ChatGPT use in undergraduate chemistry laboratories. *Journal of Chemical Education, 100*(4), 1434–1436. https://doi.org/10.1021/acs.jchemed.3c00006

Jones, R. M., Rudler, E.-M., & Preston, C. (2024). Exploring the concept of valence and the nature of science via generative artificial intelligence and general chemistry textbooks. *Journal of Chemical Education, 101*(8), 3276–3283. https://doi.org/10.1021/acs.jchemed.4c00271

Latendresse, M., Malerich, J. P., Herson, J., Krummenacker, M., Szeto, J., Vu, V.-A., Collins, N., & Madrid, P. B. (2023). SynRoute: A retrosynthetic planning software. *Journal of Chemical Information and Modeling, 63*(17), 5484–5495. https://doi.org/10.1021/acs.jcim.3c00491

Lear, B. J. (2024). Using ChatGPT-4 to teach the design of data visualizations. *Journal of Chemical Education, 101*(7), 2749–2756. https://doi.org/10.1021/acs.jchemed.4c00228

Leon, A. J., & Vidhani, D. (2023). ChatGPT needs a chemistry tutor too. *Journal of Chemical Education, 100*(10), 3859–3865. https://doi.org/10.1021/acs.jchemed.3c00288

Li, M., Enkhtur, A., Cheng, F., & Yamamoto, B. A. (2023). Ethical implications of ChatGPT in higher education: A scoping review. *arXiv preprint* arXiv:2311.14378

Lim, W. M., Gunasekara, A., Pallant, J. L., Pallant, J. I., & Pechenkina, E. (2023). Generative AI and the future of education: Ragnarök or reformation? A paradoxical perspective from management educators. *The International Journal of Management Education, 21*(2), 100790. https://doi.org/10.1016/j.ijme.2023.100790

Mahapatra, S. (2024). Impact of ChatGPT on ESL students' academic writing skills: A mixed methods intervention study. *Smart Learning Environments, 11*(1), 9. https://doi.org/10.1186/s40561-024-00295-9

Mendez, J. D. (2024). Student perceptions of artificial intelligence utility in the introductory chemistry classroom. *Journal of Chemical Education, 101*(8), 3547–3549. https://doi.org/10.1021/acs.jchemed.4c00075

Mishra, P., & Koehler, M. J. (2006). Technological pedagogical content knowledge: A framework for teacher knowledge. *Teachers College Record, 108*(6), 1017–1054. https://doi.org/10.1111/j.1467-9620.2006.00684.x

Moore, E. B., Chamberlain, J. M., Parson, R., & Perkins, K. K. (2014). PhET interactive simulations: Transformative tools for teaching chemistry. *Journal of Chemical Education, 91*(8), 1191–1197. https://doi.org/10.1021/ed4005084

Pence, H. E., Hightower, G., Forlenza, J., Leonard, K., McLellan, A., Suero, A., Amoah, B., Mbow, M., Borner, S., Castillo, A., & Pence, L. E. (2024). Using generative AI systems for critical thinking engagement in an advanced chemistry course: A case study. *Journal of Chemical Education, 101*(9), 3789–3794. https://doi.org/10.1021/acs.jchemed.4c00242

Reddy, M. R., Walter, N. G., & Sevryugina, Y. V. (2024). Implementation and evaluation of a ChatGPT-assisted special topics writing assignment in biochemistry. *Journal of Chemical Education, 101*(7), 2740–2748. https://doi.org/10.1021/acs.jchemed.4c00226

Rojas, A. J. (2024). An investigation into ChatGPT's application for a scientific writing assignment. *Journal of Chemical Education, 101*(5), 1959–1965. https://doi.org/10.1021/acs.jchemed.4c00034

Rudolph, J., Tan, S., & Tan, S. (2023). ChatGPT: Bullshit spewer or the end of traditional assessments in higher education? *Journal of Applied Learning and Teaching, 6*(1), 342–363.

Ruff, E. F., Engen, M. A., Franz, J. L., Mauser, J. F., West, J. K., & Zemke, J. M. O. (2024). ChatGPT writing assistance and evaluation assignments across the chemistry curriculum. *Journal of Chemical Education, 101*(6), 2483–2492. https://doi.org/10.1021/acs.jchemed.4c00248

Ruff, E. F., & Zemke, J. M. O. (2025). Discussing the ethics of professional AI use in undergraduate chemistry courses. *Journal of Chemical Education, 102*(4), 1457–1464. https://doi.org/10.1021/acs.jchemed.4c01038

Ryan, B. J. (2019). Integration of technology in the chemistry classroom and laboratory. In M. Seery & C. M. Donnell (Eds.), *Teaching chemistry in higher education: A Festschrift in Honour of Professor Tina Overton* (pp. 39–54). Creathach Press. https://books.google.no/books?id=lZjFDwAAQBAJ

Schrier, J. (2024). Comment on "comparing the performance of college chemistry students with ChatGPT for calculations involving acids and bases." *Journal of Chemical Education, 101*(5), 1782–1784. https://doi.org/10.1021/acs.jchemed.4c00058

Sorenson, B., & Hanson, K. (2024). Identifying generative artificial intelligence Chatbot use on multiple-choice, general chemistry exams using Rasch analysis. *Journal of Chemical Education, 101*(8), 3216–3223. https://doi.org/10.1021/acs.jchemed.4c00165

Trčková, K., Maršálek, R., Žáček, M., & Teplá, M. (2024). Why do fishes die? Lab assignment created using ChatGPT. *Journal of Chemical Education, 101*(8), 3171–3178. https://doi.org/10.1021/acs.jchemed.4c00130

Tuvi-Arad, I., & Blonder, R. (2019). Technology in the service of pedagogy: Teaching with chemistry databases. *Israel Journal of Chemistry, 59*(6–7), 572–582. https://doi.org/10.1002/ijch.201800076

Tyson, J. (2023). Shortcomings of ChatGPT. *Journal of Chemical Education, 100*(8), 3098–3101. https://doi.org/10.1021/acs.jchemed.3c00361

West, J. K., Franz, J. L., Hein, S. M., Leverentz-Culp, H. R., Mauser, J. F., Ruff, E. F., & Zemke, J. M. (2023). An analysis of AI-generated laboratory reports across the chemistry curriculum and student perceptions of ChatGPT. *Journal of Chemical Education, 100*(11), 4351–4359. https://doi.org/10.1021/acs.jchemed.3c00581

Yaron, D., Karabinos, M., Lange, D., Greeno, J. G., & Leinhardt, G. (2010). The ChemCollective—Virtual labs for introductory chemistry courses. *Science, 328*(5978), 584–585. https://doi.org/10.1126/science.1182435

Yik, B. J., & Dood, A. J. (2024). ChatGPT convincingly explains organic chemistry reaction mechanisms slightly inaccurately with high levels of explanation sophistication. *Journal of Chemical Education, 101*(5), 1836–1846. https://doi.org/10.1021/acs.jchemed.4c00235

Young, J. D., Dawood, L., & Lewis, S. E. (2024). Chemistry students' artificial intelligence literacy through their critical reflections of Chatbot responses. *Journal of Chemical Education, 101*(6), 2466–2474. https://doi.org/10.1021/acs.jchemed.4c00154

Yuriev, E., Wink, D. J., & Holme, T. A. (2023). Virtual special issue call for papers: Investigating the uses and impacts of generative artificial intelligence in chemistry education. *Journal of Chemical Education, 100*(9), 3168–3170. https://doi.org/10.1021/acs.jchemed.3c00829

Yuriev, E., Wink, D. J., & Holme, T. A. (2024). The dawn of generative artificial intelligence in chemistry education. *Journal of Chemical Education, 101*(8), 2957–2959. https://doi.org/10.1021/acs.jchemed.4c00836

9 Teaching and Learning Chemistry Outdoors

Abstract

This chapter considers chemistry teaching and learning outside of the classroom, the potential benefits of moving parts of chemistry lessons outdoors, and what teacher preparations and competencies are relevant to successfully make this happen.

The practice of taking students outdoors on field trips, excursions, camping, to a nearby school yard, to game parks, nearby industries (factories), to science centers, gardens, zoos, amusement parks, playgrounds, etc., is well known and well-established in schools. Depending on how schools are structured, teachers of any subject can organize an outdoor activity for their students, either with a formal and clear pre-determined learning goal or informally to simply give their students the opportunity to connect their learning with phenomena in everyday life. In some science fields like biology and environmental sciences, taking students outside the classroom and performing curricula related activities outside of the classroom is routinized. Student excursions are, for example, common in biology classrooms where students being guided by their teachers set out to explore, observe, and collect data about biological processes, organisms and ecosystems in real contexts. It is also possible for teachers of different subjects to arrange field work and camping activities with pre-organized learning activities. *Outdoor education* is the term that has been used in education to coin all these kinds of school-related academic activities which are organized and carried out outside the formal classroom (or school laboratory for science disciplines) (Eaton, 2000; Gilbertson et al., 2022; Jim et al., 2014). Outdoor education is an encompassing term that does not only include activities taking place outside the classroom in the open (playground, field, garden, etc.), but also those activities taking place inside closed structures (e.g., science museums, science centers, etc.) which are not classrooms or school laboratories.

F. Kayima, *Preparing for Chemistry Teaching*, Springer Texts in Education,
https://doi.org/10.1007/978-3-031-97122-8_9

Three commonly known traditional forms of outdoor activities are distinguished in educational literature, namely, (i) excursions, (ii) field work and (iii) school camps or camping (Andersen & Fiskum, 2014). Excursions are considered short trips organized and controlled by the teacher. The teacher normally has full control over all planned activities and students are mainly observers unless told otherwise. Excursion trips are short, lasting from a few hours to a full day. It could for example be a visit to a museum, zoo, science center, to a historical site, to a local food processing industry or to a theater. The goal is often to provide students with experiences to complement what they have learnt in classrooms. Fieldwork on the other hand is a more subject focused form of outdoor education. It is tightly connected to one or more learning goals being addressed in the classroom. For example, the teacher can plan a fieldwork activity with a concrete problem statement, where students will explore a given phenomenon, collect data in a natural environment, and use the collected data to respond to the initial problem statement having returned from the field. The duration of fieldwork activities can range from a few hours to several days depending on the nature of the activity. For it to be a success in terms of achieving the intended goal of the field trip, teachers are expected to have a good plan for fieldwork activities that considers the work before and after the field trip. The work before the trip to the field could involve for example, deciding upon the topic of investigation and guiding (research) questions, the knowledge and skills students will need to be able to address the set-out investigation tasks, etc. The teacher can also assess the relevant tools and resources students will need once in the field and make these available depending on the number of degrees of freedom provided in the task to be explored, that's whether the task is closed with a working procedure provided or whether its open for students to explore different alternatives to solving the task, etc.

School camp (or camping) is another form of outdoor activity. It is an organized trip lasting several days (normally 2 to 5 days) with well pre-planned learning activities. The activities could be a mix of subject specific learning activities, outdoor social activities, and activities geared towards personal development. For example, a chemistry teacher can organize with support from the school administration and the local community a three-day school camp with a theme, "Plastics along a riverbank". S/he can plan to camp with students along or near a riverbank, with concrete planned activities to be carried out during the three days. Depending on the curriculum, the teacher can decide what discipline specific aspects will be explored with hands-on experiences, and what social-related activities will be possible depending on the nature of the camping site. In some countries, there are already organized non-profit organizations which organize and offer school camping to children of different ages, covering and addressing different societal and environmental -related themes. In such cases, the teacher's task is to assess what activities provide learning experiences that relate to the teaching that goes on in classrooms, that the teacher wishes his/her students to have.

9.1 Chemistry Outdoors

Generally, the benefits of outdoor learning experiences are well-documented, including motivating students, raising students' awareness about advances in science and technological applications, developing and enhancing transferrable skills such as data collection, presentation and observational skills, etc., bridging classroom instruction with everyday natural phenomena, eliciting perceptual changes and enhancing and developing environmentally friendly practices, etc. Indeed, there is growing evidence as indicated in several research reports (e.g.; Arıkan, 2023; Ayotte-Beaudet et al., 2017; Ceci, 2015; Engl & Risch, 2016; Guerrero & Reiss, 2020; Haraldsson et al., 2024; Höper, 2017; Höper & Köller, 2018; Höper et al., 2022; Maciejowska, 2007; Mirrahimi et al., 2011; Remmen et al., 2021; Stevenson et al., 2021) that outdoor learning experiences enhance students' cognitive, affective and social competences. Moreover, it is one thing to provide outdoor learning experiences for students to connect and relate what they learn in classrooms to phenomena in everyday life by exploiting any of the several natural settings mentioned above (excursions, field trips, visits to nearly school yards, gardens, zoos, etc.), and another to teach chemistry concepts outside of the four walls of the classroom or the laboratory. Our interest is the latter. We want to take part of the chemistry lessons planned from the curriculum and facilitate students' learning of these concepts outside of the classroom or the laboratory. Well planned and executed outdoor chemistry lessons can reinforce students understanding of basic chemical concepts and principles (Borrows, 2006; Höper, 2025; Höper et al., 2022). This is what this chapter is about, the possibility of learning chemistry content knowledge outdoors. It is about the teacher being able to identify chemistry aspects in the curriculum, develop and implement an outdoor teaching activity covering these specific chemistry concepts.

9.1.1 Planning Chemistry Outdoors

The idea of learning chemistry outside of the classroom can be informed by at least two theoretical perspectives, namely (i) the context-based teaching and (ii) the situated learning perspective.

9.1.1.1 A Context-Based Approach to Teaching

A context-based approach is a well-known teaching approach in chemistry education even though several research reports indicate that most teachers and educators who employ this teaching approach do it inside classrooms (e.g.; Höper et al., 2022; Overton et al., 2013). A context-based teaching and learning approach considers the "use of contexts and applications of science as a starting point for developing scientific understanding" (Bennett et al., 2005, p. 1523). Contexts for applications of science range from the social, economic, technological, industrial, and environmental contexts (Bennett & Holman, 2002; Overton, 2007; Rose,

2012). A context-based teaching entails a teacher selecting any of the above contexts that s/he considers relevant for the target students (and learning), for example contexts closest to students' immediate lives, e.g., food, clothing, sports, etc., and then identifying the science content that students would engage with in interaction with the selected contexts (Bennett & Holman, 2002; Bennett et al., 2005). The science content aspects to be learned are introduced to the students during that moment they are needed to support students to understand the features of the context they are exploring. This does not imply that students get introduced to new science concepts for the first time during the time they are engaged and interacting with features of the selected context of interest. The teacher has, in one way or another, introduced these or parts of these concepts in the classroom before the outdoor expedition as the basis knowledge to be applied during the outdoor expedition. What is implied here is that science knowledge concepts are invoked and applied as it is needed in those instances when the outdoor activity is ongoing. Bennett et al. (2005) called this introducing scientific ideas on a "need to know" basis (p. 1523). In other words, science contents are the knowledge base the students need to fill in the gaps as they explore the contexts of interest.

If we are planning to employ a context-based approach as described above and take chemistry outdoors, we should be clear about the fact that not all chemical concepts can be learned or taught outdoors. For example, how can we teach topics like the dual wave-particle nature of the electron, chemical equilibrium, redox, and many others by exclusively using a context-based approach, and moreover by taking chemistry outdoors? This is definitely a very complex mission for the teacher to take on. What is most likely and plausible is indeed taking only parts of the chemistry teaching outdoors. Chemistry outdoors is more likely to succeed as a compliment to classroom teaching than as an independent and self-standing. This is where the idea of introducing chemistry concepts on "need to know" basis makes sense. Students need a knowledge base with which they can explore and make sense of the features of the contexts the teacher has selected to be explored. This knowledge base can either be provided in classrooms before setting out to explore the target features of the selected contexts, or the teacher can plan in such a way that students have the necessary resources to help them find the relevant science concepts to help them piece together and understand different elements of the explored phenomena (features of the explored contexts).

Several educators have examined the possibility of taking parts of chemistry outdoors with some examples that can be drawn on. One such example is by Höper and Köller (2018), a study where the authors explore the potential learning benefits of taking student teachers out of the lab into nature. In this study, students act as "molecule detectives", equipped with a "chemistry tool bag" and then proceed to the field (nature). The bag contains "easy to use" testing kits for glucose, and starch, and the task is to find carbohydrates in the near environment (surrounding). Students are provided with clear procedural guidelines to be followed to complete the task successfully. This outdoor activity is simply a part of the teaching unit on glucose and other carbohydrates which is covered in a series of lessons. The authors note that the outdoor activity enabled students to make connections between content knowledge and practical aspects in natural settings, for example, enhanced understanding of solubility and concentration concepts. In the second example, Höper et al. (2022) describe a teaching unit where student teachers identify and examine corroded metals in the nearby outdoor environment. The underlying goal for the teaching unit according to the authors was to support students' understanding of corrosion. By exploring corroded metals, students were to elaborate and apply their basic knowledge of redox chemistry. The entire unit comprised three parts, first in the classroom where students were introduced to the relevant knowledge base (chemistry concepts about corrosion and metal surfaces), the second part—the outdoor session where students, equipped with necessary tools and resources depending on the nature of the task, set out to explore and complete the task, and then the third part- where students are back in the classroom upon completing the outdoor expedition. Both teacher and students already have a clear plan with tasks and responsibilities about what they are supposed to do in classroom before the outdoor expedition, during the outdoor expedition and after when they return to classroom. Again, this second example is like the first one in the sense that students are provided with a knowledge base before the outdoor activity, along with the necessary equipment and tools to accomplish the assigned outdoor activity. A knowledge base is needed to be able to interact with and understand the features of the explored contexts as earlier mentioned. Moreover, the outdoor activity is contextualized and located within the main learning unit—that is about redox chemistry, which stretches over a series of lessons conducted inside the classroom or the laboratory.

For the third example, Borrows (2006) discusses some examples of the kind of chemistry aspects that can be explored by considering and examining objects in the nearby surroundings such as concrete buildings and churchyards. Two case examples from Borrows' report include one possibility where the teacher and students can consider the chemical composition of cement, starting with exploring how cement is made, what the constituent raw materials are, and whether students can relate these raw materials (calcium carbonate (limestone) and clay) with what they know from classroom teaching, identifying familiar reactions that they can relate to. Borrows highlights for example the thermal decomposition of calcium carbonate to produce calcium oxide and carbon dioxide gas, which is an endothermic reaction and then its reverse reaction that can occur for example at room temperature. The teacher can choose to guide the discussion about reversible reactions or to consider the presence of calcium oxide in cement as responsible for corrosiveness, etc. The second possibility addressed by Barrows involves students exploring a churchyard together with the teacher, examining and identifying different aspects of chemical interest, including identifying different metals/elements, nature, and properties, exploration of rusting as an example of a redox reaction, etc.

9.1.1.2 The Situated Learning Perspective

A situated learning perspective emphasizes the situatedness of learners in particular contexts or environments (Sadler, 2009). It entails that knowledge and skills are acquired in "contexts that reflect how knowledge is obtained and applied in everyday situations." (Stein, 1998, p. 1). The contexts or environments within which learners are located and are participating in, are pertinent to their learning. These either afford (facilitate), or constrain learning and it is in these contexts or environments that "learning and knowing takes place (Sadler, 2009). Stein (1998) summarizes four premises of situated learning, ideas originally presented by Anderson et al. (1996) and Wilson (1993), which can inform the planning and developing of classroom activities, namely (i) learning is grounded in everyday situations, (ii) knowledge is acquired situationally and transfers only to similar situations, (iii) learning results from a social process and includes engagement with other skills and attributes (thinking, perceiving, problem-solving, interactions, etc.), in addition to declarative and procedural knowledge, and (iv) learning is not separated from the world of action, exists in robust, complex, social environments made up of actors, actions and situations (Stein, 1998, p. 3). Based on these four premises underlying situated learning, Stein (1998) goes on to argue that according to situated learning, students learn content knowledge through engagement in activities as opposed to acquiring knowledge in discrete packages from their instructors. This argument is also supported by other science educators embracing a situated learning theoretical perspective that considers knowing and learning as participation in environments and engagement with communities constituting these environments (e.g.:, Chaiklin & Lave, 1996; Greeno et al., 2013; Sadler, 2009).

9.1.2 A Context-Based Approach Versus a Situated Learning Perspective

Both the context-based approach and the situated learning perspective revolve around teachers selecting relevant contexts for application of science and then use these contexts as the basis for understanding science content and its applications. Moreover, both perspectives hinge on the importance of students' active participation and hence have a strong underpinning in the work of Dewey, Vygotsky and others behind a (socio) constructivist's thinking about teaching and learning. However, situated learning as a learning perspective goes beyond simply selecting a context and exploring certain features of the selected context. It demands that students are emersed in authentic contexts, be part of a community performing the activities of interest and learn from actively participating in the real activities being completed or performed by that community the students have decided to be part of. Sadler (2009) gives a hypothetical example of what participation in an authentic context looks like. He describes a student who receives a summer opportunity to work in the laboratory as a practicing student. The student gets introduced to the lab and to the entire team she will be working with, who have a range of different expertise. The student is then assigned well-defined tasks directly affecting an ongoing project, for example, involving "genetic analyses including frequent use of PCR". At the start, she gets some assistance from an assigned colleague who walks her through the PCR protocol and demonstrates the process.

The student then gradually starts to work independently in the weeks that follow, including assuming new responsibilities as her/his knowledge and skills improve and as needs arise within the project. That is,

> She starts by helping to set up new crosses in the *Drosophila* lines that are the foci of her project, extracts new DNA samples from the resulting flies and runs some sequencing gels. Along the way, she very likely made mistakes (killed a few flies, contaminated a few DNA samples or left too many air bubbles in a gel), asked questions of anyone who would listen and struggled to make sense of what she was doing. By the end of the summer, she has gained entirely new perspectives on *Drosophila*, DNA, PCR and a range of other physical and conceptual tools that she used to negotiate this experience. The student learned about science and genetics through participation in a community of researchers working on authentic problems related to genetics. (Sadler, 2009, p. 3)

Planning a teaching experience similar to that described by Sadler (2009) above may not be possible all the time, especially if we want to apply situated learning ideas to outdoor chemistry. It is reasonably not possible to get placements for all students in a chemistry class, notwithstanding the resources that might be needed to achieve this. Nevertheless, Sadler's example gives us a good example of learning as participation and hence an opportunity for a chemistry teacher who wishes to plan a small expedition where students can learn chemistry while participating in real community activities. Alternatively, Sadler et al. (2011) explored the possibility of learning science content by anchoring the teaching to a socio-scientific issue (ISS). The authors developed what they called an "ISS-based curriculum and instruction" project, where they selected an ISS issue, developed both teaching and assessment materials, and carried out an inside classroom teaching using the developed materials. The teaching plan that stretched over a 3-weeks period, comprised nine interconnected lessons, starting with an introduction to global climate change (the central ISS issue), exploration of different views about climate change from politicians, and other interest groups, four lessons focused on understanding of the science underlying global climate change covering themes like—the particulate nature of gases, climate and temperature, atmospheric composition, combustion, greenhouse gases, greenhouse effect, composition of Ozone and ozone layer depletion process and reactions, etc., and other lessons aimed at developing students' analytical, observational, socio-scientific reasoning, critical thinking skills, etc. Although the authors implemented their teaching plan inside the classroom, an ISS -based instructional plan opens for teaching chemistry outdoors, and the fundamental understanding behind its organization is in line with the idea of selecting contexts for science applications and using these to teach science content. Indeed, a similar teaching plan like that described by Sadler et al. (2011) can be planned with a different ISS issue with its corresponding science content. It is then possible to take parts of the teaching outdoors depending on the availability of resources, including the teacher's competence.

9.2 Chemistry Outdoors Should be About Chemistry

In this chapter, emphasis has been placed on discussing two theoretical perspectives to give you a fundamental basis for planning a chemistry teaching outdoors. It is often easy to simply plan to take your students outdoors for the sake of taking them. After all, you had the same experience as a student, of going on field trips, visiting museums, science parks, etc., without learning something concrete. You do not wish to waste yours and students' time on expeditions whose value in terms of learning cannot be concretized, documented, and later alone, valued and visualized by the learners themselves. It is possible to teach chemistry outdoors, but you need to have the relevant knowledge and competences to do so. This chapter provides a didactic knowledge entrance through an elucidation of the relevant theoretical perspectives supplemented by concrete examples from educators who have applied these perspectives in classrooms. This didactic entrance can inform how you can plan and structure your outdoor teaching plan. However, you will need the chemistry content knowledge. Moreover, it is not only the chemistry content knowledge presented in textbooks that you need, but knowledge of this chemistry content knowledge located in real contexts where you will plan to take your students.

Take an example of the concrete -cement relationship (Borrows, 2006) that we discussed earlier or the corrosion investigation expedition (Höper et al., 2022). In both cases, the teacher needs to know what is there in nature to be found about chemistry, that which relates to what is happening in classrooms. In Borrows' example, the teacher is able to take students through the process of making cement, what chemical compounds are involved, and the conditions for the reactions to take place, etc. He is able to identify the relevant reactions that students will be able to relate to from their classroom learning, while at the same time avoiding the danger of introducing other complex reactions which would negatively affect students' motivation going forward. It implies the teacher has done enough reading to understand the composition of cement and concrete to be able to develop a plan that suits the level of the learners and to give learners the opportunity to appreciate the use of cement and concrete as examples they associate with their everyday life. In other words, while there is a lot of chemistry around us, this chemistry is hidden in the everyday language and interwoven in our daily life and practices. This everyday language is different from the language of chemistry, the medium we use to communicate about chemistry in classrooms. Consequently, it is you the teacher to help students to uncover this chemistry around us by helping to bridge the gap between these two languages. You cannot do this without you yourself understanding what kind of chemistry is out there for you and your students to explore and connect to the theoretical chemistry as discussed in classrooms.

References

Andersen, H. P., & Fiskum, T. A. (2014). Hva er uteskole?-Noen begrepsavklaringer. In T. A. Fiskum & J. A. Husby (Eds.), *Uteskoledidaktikk. Ta fagene med ut* (pp. 15–29). Cappelen Damm.

Anderson, J. R., Reder, L. M., & Simon, H. A. (1996). Situated learning and education. *Educational Researcher, 25*(4), 5–11.

Arıkan, K. (2023). A comparison of indoor and outdoor biology education: What is the effect on student knowledge, attitudes, and retention? *Journal of Biological Education, 57*(4), 727–745. https://doi.org/10.1080/00219266.2021.1950809

Ayotte-Beaudet, J.-P., Potvin, P., Lapierre, H. G., & Glackin, M. (2017). Teaching and learning science outdoors in schools' immediate surroundings at K-12 levels: A meta-synthesis. *Eurasia Journal of Mathematics, Science and Technology Education, 13*(8), 5343–5363.

Bennett, J., Grasel, C., Parchmann, I., & Waddington, D. (2005). Context-based and conventional approaches to teaching chemistry: Comparing teachers' views. *International Journal of Science Education, 27*(13), 1521–1547.

Bennett, J., & Holman, J. (2002). Context-based approaches to the teaching of chemistry: What are they and what are their effects? In J. K. Gilbert, O. Jong, R. Justi, D. F. Treagust, & J. H. Driel (Eds.), *Chemical education: Towards research-based practice* (pp. 165–184). Kluwer Academic Publishers.

Borrows, P. (2006). Chemistry outdoors. *School Science Review, 87*(320), 23.

Ceci, C. (2015). Take concepts of chemistry out of the classroom. *Nature, 522*(7554), 7–7. https://doi.org/10.1038/522007a

Chaiklin, S., & Lave, J. (1996). *Understanding practice: Perspectives on activity and context.* Cambridge University Press. https://books.google.no/books?id=VaNl678-E0EC

Eaton, D. (2000). *Cognitive and affective learning in outdoor education* University of Toronto.

Engl, A., & Risch, B. (2016). Natural chemistry-outdoors. *Green Teacher, 109*(1), 39–42.

Gilbertson, K., Ewert, A., Siklander, P., & Bates, T. (2022). *Outdoor education: Methods and strategies.* Human Kinetics.

Greeno, J. G., Collins, A. M., & Resnick, L. B. (2013). Cognition and Learning. In D. C. Berliner & R. C. Calfee (Eds.), *Handbook of educational psychology.* Routledge. https://books.google.no/books?id=yw8FmsvgzuQC

Guerrero, G. R., & Reiss, M. J. (2020). Science outside the classroom: Exploring opportunities from interdisciplinarity and research–practice partnerships. *International Journal of Science Education, 42*(9), 1522–1543. https://doi.org/10.1080/09500693.2020.1767317

Haraldsson, K., Göranson, M., & Lindgren, E.-C. (2024). "It is easier to learn when you are out": An ethnographic study of teaching science subjects through outdoor learning at compulsory school. *Journal of Outdoor and Environmental Education.* https://doi.org/10.1007/s42322-024-00172-6

Höper, J. (2017). *Natural experiments: taking the lab outdoors.*

Höper, J. (2025). Mind the gap between chemistry and outdoor education: Primary and lower secondary preservice teachers' beliefs during an outdoor-based introduction to chemistry in science education. *Journal of Science Teacher Education, 36*(3), 370–396. https://doi.org/10.1080/1046560X.2024.2406661

Höper, J., Jegstad, K. M., & Remmen, K. B. (2022). Student teachers' problem-based investigations of chemical phenomena in the nearby outdoor environment. *Chemistry Education Research and Practice, 23*(2), 361–372. https://doi.org/10.1039/d1rp00127b

Höper, J., & Köller, H.-G. (2018). Outdoor chemistry in teacher education—A case study about finding carbohydrates in nature. *LUMAT: International Journal on Math, Science and Technology Education, 6*(2), 27–45. https://doi.org/10.31129/LUMAT.6.2.314

Jim, S., Melissa, P., & Maura, C. (2014). Outdoor education. In C. P. G. S. U. A. G. J. Lund, J. Lund, & D. Tannehill (Eds.), *Standards-based physical education curriculum development.* Jones & Bartlett Learning. https://books.google.no/books?id=HYmOAwAAQBAJ

Maciejowska, I. (2007). Chemical education outside a classroom–step by step. *Natural Science Education, 3*(20), 51–58.

Mirrahimi, S., Tawil, N. M., Abdullah, N. A. G., Surat, M., & Usman, I. M. S. (2011). Developing conducive sustainable outdoor learning: the impact of natural environment on learning, social and emotional intelligence. *Procedia Engineering, 20*, 389–396. https://doi.org/10.1016/j.proeng.2011.11.181

Overton, T. (2007). Context and problem-based learning. *New Directions in the Teaching of Natural Sciences* (3), 7–12.

Overton, T., Potter, N., & Leng, C. (2013). A study of approaches to solving open-ended problems in chemistry. *Chemistry Education Research and Practice, 14*(4), 468–475.

Remmen, K. B., Jegstad, K. M., & Höper, J. (2021). Preservice teachers' reflections on outdoor science activities following an outdoor chemistry unit. *Journal of Science Teacher Education, 32*(4), 425–443. https://doi.org/10.1080/1046560X.2020.1847967

Rose, D. E. (2012). Context-based learning. In N. M. Seel (Ed.), *Encyclopedia of the sciences of learning* (pp. 799–802). Springer US. https://doi.org/10.1007/978-1-4419-1428-6_1872

Sadler, T. D. (2009). Situated learning in science education: Socio-scientific issues as contexts for practice. *Studies in Science Education, 45*(1), 1–42. https://doi.org/10.1080/03057260802681839

Sadler, T. D., Klosterman, M. L., & Topcu, M. S. (2011). Learning science content and socio-scientific reasoning through classroom explorations of global climate change. In *Socio-scientific issues in the classroom: Teaching, learning and research* (pp. 45–77). Springer.

Stein, D. (1998). Situated learning in adult education. ERIC Digest No. 195. In *ERIC, Resources in Education (RIE)* (pp. 1–4). ERIC Clearinghouse on Adult, Career, and Vocational Education, Columbus, OH. https://www.proquest.com/other-sources/situated-learning-adult-education-eric-digest-no/docview/62542070/se-2?accountid=12870

Stevenson, K. T., Szczytko, R. E., Carrier, S. J., & Peterson, M. N. (2021). How outdoor science education can help girls stay engaged with science. *International Journal of Science Education, 43*(7), 1090–1111. https://doi.org/10.1080/09500693.2021.1900948

Wilson, A. L. (1993). The promise of situated cognition. *New Directions for Adult and Continuing Education, 1993*(57), 71–79. https://doi.org/10.1002/ace.36719935709

Integrating Sustainability Issues in Chemistry

10

Abstract

Sustainability concerns the possibility for both humankind and other diverse life forms sharing this planet (Earth) to continue living on the planet and using its resources within the limits afforded by the planet. This chapter considers and presents possible methodological frameworks and pedagogical approaches (and models) with which teachers can integrate sustainability issues in chemistry.

In this chapter, we turn attention to Sustainable Development (SD) and how teachers can approach the teaching and learning of chemistry in ways that promote sustainable development (SD) efforts. Sustainability concerns the possibility for both humankind and other diverse life forms sharing this planet (Earth) to continue living on the planet and using its resources within the limits afforded by the planet. The need for sustainability in all human activity and actions comes as a response to the many problems and challenges that face our planet. Our continued exploitation of the planet's resources and our way of life (economic, social and environmental human activity) has had consequential and serious effects, to the point that the very survival of mankind on the planet is increasingly threatened. The society's environmental footprints are too damaging that urgent action is needed to ensure that we can live sustainably within the planet's boundaries (Rockström et al., 2009; Steffen et al., 2015). Sustainable Development (SD) is envisioned as a pathway, a journey or a process towards achieving sustainability (Lozano, 2008), and it is concerned with streamlining human activities and actions (development) such that while we simultaneously pursue economic prosperity, environmental quality, and social equity, these pursuits are made in the service of both the current and future generations.

Chemistry and chemistry education have a central role in achieving sustainability, first by providing the necessary basis for understanding and addressing

F. Kayima, *Preparing for Chemistry Teaching*, Springer Texts in Education,
https://doi.org/10.1007/978-3-031-97122-8_10

current and future sustainable challenges, and second, by contributing to solutions to ensure sustainable living within planet's boundaries (Mahaffy et al., 2019; Wissinger et al., 2021). Moreover, the nature of chemistry and the chemistry enterprise categorically demands that chemistry and chemistry education are at the forefront in developing and promoting mechanisms and solutions to deal with sustainability problems. Chemistry comprises the material and a basis for energy that drives our economies and society. Many products we use in our daily lives are either based on chemistry or are chemical products themselves. Chemistry is embedded in our everyday lives, in the foods we eat, in drinks and beverages, the cleaning detergents, softeners, and disinfectants we use, in medicine and health products, clothing, food colors and personal care products, in electronics and agricultural production and other chemical technologies, to mention but a few. As such, any efforts to achieve sustainability cannot be imagined without contributions from chemistry and chemistry education (Anastas, 2003; Anastas & Zimmerman, 2018).

Indeed, the inception of green chemistry (GC) in the 1980s and sustainable chemistry (SC) in the 1990s was and is a response to the demand for sustainability in the chemistry enterprise, including in chemistry research, in production and other related industries (Eilks & Linkwitz, 2022; Juntunen & Aksela, 2014; Zuin et al., 2021). GC was and still viewed as an approach that "fulfills a fundamental and crosscutting role that is essential to the critical pathway toward sustainability" (Anastas, 2003; Idul & Walag, 2024; Timmer et al., 2018). It is "the utilization of a set of principles that reduces or eliminates the use or generation of hazardous substances in the design, manufacture and application of chemical products." (Anastas & Warner, 1998; Idul & Walag, 2024). SC on the other hand, is " a scientific concept that seeks to improve the efficiency with which natural resources are used to meet human needs for chemical products and services."(Zuin et al., 2021, p. 1596).

The central understanding and aims of both green chemistry (GC) and sustainable chemistry (SC) largely overlap despite emerging from different origins and traditions (Eilks & Linkwitz, 2022). Horváth (2018) expresses that SC should be concerned with the use of resources that allows a natural process of replacing them, and a rate of waste generation that is not faster than their remediation. Moreover, he lays bare the fact that "not all sustainable chemicals or reactions or processes could be green", as it is that "some parts of green chemistry may not be sustainable at all" (p. 369). As such, target developments and innovations within green and sustainable chemistry, according to Horváth (2018) would preferably consider the selection of those chemicals, reactions, syntheses routes, pathways and processes, which are sustainable and green at the same time. Blum et al. (2017) proposed a more encompassing definition of sustainable chemistry that goes beyond green chemistry and the operational safe use of chemicals to include social, economic and policy aspects. They consider SC to be a holistic approach that considers "the entire lifecycle of chemical products and the related system of actors, institutions and culture" (p.94). SC thus according to Blum et al. (2017) considers stakeholders and their responsibilities along the life cycle chain of chemicals, ensuring that conditions relating to health, environmental, social, research, science and economics, are considered and balanced within the capacity-limits of the planet.

10.1 Contributing to Sustainability Efforts Through Chemistry Teaching

As it is for all teachers, chemistry teachers also have a role and are expected to contribute to achieving sustainability through the work of teaching. This means having or acquiring the relevant didactic competences to be able to plan and execute chemistry lessons with sustainability issues integrated. It implies acquiring a good understanding about the need for sustainability and how chemistry can contribute to sustainable development at the teaching and learning level. We should have the awareness that if we simply select chemical theories, concepts, models, and laws as we know them to appear in chemistry textbooks and teach these to our students, the level of students' understanding of sustainability issues and what their current and future obligations are, will be limited. The broad consensus therefore is that integration of sustainability issues in chemistry teaching and learning should be more prominent (e.g.; Aoki et al., 2022; Armstrong et al., 2024; Azizah & Yonata, 2023; Burmeister et al., 2012; Idul & Walag, 2024; Mahaffy et al., 2019; Paristiowati et al., 2022; Timmer et al., 2018; Wissinger et al., 2021). To integrate sustainability issues in our chemistry teaching means planning to teach in a different way. This is what this chapter is about. How can chemistry teachers plan to teach chemistry in ways that take up and address sustainable development issues? Or with what teaching methods, didactic or pedagogical frameworks can chemistry teachers integrate sustainability issues into chemistry teaching and learning?

There are several ways the above question has been addressed by education researchers and educators in education generally and in chemistry education research. Some suggestions from chemistry education research include a holistic approach to ESD in chemistry education (Juntunen & Aksela, 2014), an Elliptic model for applying ESD in chemistry teaching and learning (Jegstad & Sinnes, 2015), situating SD within chemistry through systems thinking approach (Eaton et al., 2019; Mahaffy et al., 2019; Talanquer & Szozda, 2024; Wissinger et al., 2021; York & Orgill, 2020), an interdisciplinary approach—teaching chemistry concepts interposed with other disciplines (Iyere, 2008), use of project-based (or problem-based) learning approach to sustainability integration into chemistry (e.g.; Azizah & Yonata, 2023; Cáceres-Jensen et al., 2021; Paristiowati et al., 2022), use of a process oriented guided inquiry learning—POGIL (e.g.,; Idul et al., 2025), the four models for integrating sustainability issues in chemistry including (i) integrating practices from Green & Sustainable chemistry (GSC) into school practical work, (ii) including GSC practices as content in the chemistry curriculum, (iii) use of SSI that are related to sustainability issues and chemistry, and (iv) making sustainability a driver for school daily business in the short and long term (Burmeister et al., 2012; Eilks & Linkwitz, 2022), and the "Action Competence Approach"—an action-oriented teaching that considers four knowledge dimensions about sustainability problems (Jensen, 2004; Jensen & Schnack, 1997; Van Poeck et al., 2024).

What the above indicates is that there are different possible approaches to achieving the needed integration. Moreover, it should also be noted that varied conceptualizations, understandings and or descriptions of what constitutes the concepts sustainability, sustainable development and education for sustainable development have been put forward by different scholars. These various understandings are also reflected in the different frameworks and teaching approaches being suggested. The challenge that comes with having several alternatives to how we should integrate sustainability issues in chemistry teaching is that we must consider and evaluate the available alternatives before choosing the best alternative to serve our teaching interests. This cannot be realized without first having a good understanding of what frameworks or approaches have been suggested, proposed or developed in literature. Adopting and using one or more approaches/frameworks as those examples given in the above paragraph, would also depend on the contextual and curricula priorities and goals. We will consider a few examples of these teaching methods and frameworks in detail in the next part with a focus on how they are potentially operationalized in actual teaching.

10.2 The Four Basic Models for Integrating Sustainability Issues in Chemistry

Burmeister et al. (2012) suggested four basic models with which sustainable development issues can be approached in chemistry education in their review study where they analyzed chemistry education literature addressing sustainable development issues in chemical education. As you will later recognize, these suggestions are not different from what we know from before, thereby making it easier for you to relate them to your own experience. For model 1, the authors suggest adopting green chemistry principles to chemistry-practical work. The underlying statement here is about green chemistry principles. What are these principles? Do you have a good understanding of these principles and ideas about how these could be operationalized in laboratory teaching contexts? Having a basic understanding of the philosophy of green chemistry and its underlying principles is a prerequisite for you to be able to embrace and use model 1 as your teaching approach. In summary and in the context of teaching, green chemistry principles concern the handling of chemicals and lab work procedures, selecting and using less poisonous, less toxic or less hazardous alternatives and chemical substitutes wherever this is possible, it concerns dealing with chemicals in a responsible and safe manner to protect yourself, your students and the environment, working towards reducing consumption and creation of unnecessary wastes. Moreover, it requires that the teacher is familiar with the twelve principles of green chemistry (Anastas & Warner, 1998) to guide to his/her practice.

Idul et al. (2025) provide us with a concrete example of how a lesson based on green chemistry (GC) principles can be instructionally planned and organized. The authors report a study where they developed learning modules that incorporated GC principles, and then employed the Process-Oriented Guided Inquiry Learning (POGIL) approach to implement their developed modules. POGIL is "a student-centered, group-learning instructional strategy and philosophy that was developed through research on how students learn best." https://www.pogil.org/what-is-pogil. In a POGIL learning environment, students are actively engaged in mastering the concepts and content of a discipline; at the same time, they are developing important learning skills by working in self-managed teams on guided inquiry activities designed specifically for this purpose and environment." (Moog & Spencer, 2008, p. 3; Moog et al., 2006, p. 43). The approach anchors on a learning cycle comprising three steps, exploration, concept development (concept invention), and concept application (Idul et al., 2025). During the exploration phase, students engage in finding trends or patterns of regularity in the environment or in the data collected by the students or presented to them directly by the teacher. They generate and test hypotheses to explain or understand this information. The concept development phase involves developing or introducing new terms or concepts from the developed patterns or trends in the previous phase. During the application phase, students apply the newly discovered concept to new situations, thereby deepening their understanding (Moog & Spencer, 2008; Moog et al., 2006). The three- phase learning cycle is conceived as allowing students to explore and use discipline content to construct understanding and develop content mastery, while at the same time developing and enhancing their cognitive and affective process skills.

Idul et al. (2025) developed six modules integrating green chemistry concepts into core chemistry topics, namely, stoichiometric reactions, colligative properties, enthalpy, reaction order, chemical kinetics and catalysis. Some of the integrated green chemistry concepts included, atom economy and waste prevention, safer chemicals and processes, renewable feedstocks, design for energy efficiency, etc. The six modules each followed the POGIL three phase learning cycle described above, and activities for both students and teachers were designed to match the profile of POGIL learning cycle described in the above paragraph.

Considering the lesson instructional organization by Idul et al. (2025), two important aspects can be drawn. First, the authors started their intervention by training their participants (teachers) in the fundamental principles of green chemistry and the application of these. This concerns exactly what was mentioned earlier that having a good understanding of the philosophy of green chemistry and its underlying principles is a prerequisite for one to be able to plan a chemistry lesson that integrates these GC principles. Second, the authors employed a POGIL approach when designing their lessons (modules) and followed the POGIL learning cycle to plan the contents and activities of their respective modules. Again here, one needs a good understanding of the POGIL instructional philosophy and methodology to be able to apply this approach. I have provided a brief overview of this approach, however, to be able to apply the approach, one needs to study and understand the philosophy underlying the approach and how the approach is operationalized in practice. This is part of your assignment as a training or practicing chemistry teacher. Moreover, it is also possible to integrate GC principles in your chemistry teaching following any other methodological approach. Azizah and Yonata (2023) describe a study where certain principles of GC are implemented in chemistry teaching using common teaching methods and models such as discussion, concept map, or practical methods as well as Project Based Learning. Once you have a good understanding of the GC principles, it will be easier for you to plan a student-centered lesson that incorporates these GC principles, and it is possible to employ any of the known student-centered teaching methods and approaches to accomplish the intended learning.

Burmeister et al.'s (2012) model 2 considers sustainability issues as the content of the teaching. Chemistry topics, concepts and processes aligned with green and sustainable chemistry principles are selected and taught to students. Examples of relevant topics here suggested by Eilks and Linkwitz (2022) include students learning about bio-fuels, plastics and renewable energy sources, etc. Model 3 suggests selecting and using socio-scientific issues related to sustainability issues and chemistry and use these (at least one concrete issue) as a departure point into chemistry learning. One example would be about ethical dilemmas surrounding the exploitation of cobalt from Congo and the sustainability of natural elements (minerals) which are central to the sought sustainable and green future. That is, we need to exploit (use) natural elements (minerals) to advance our green technologies for a sustainable future, but are there enough of these relevant minerals to exploit? Is sustainability ensured? What about the manner through which these minerals are obtained, what are the ethical dilemmas surrounding human abuses in the process of acquiring these minerals that are vital to our green technologies? If properly planned, using ISS to teach chemistry provides widened learning opportunities to students, stretching from learning about the basic chemical principles to broader skills touching decision making and participation in society.

Model 4 suggests "chemistry education as a part of ESD-driven school development" (Burmeister et al., 2012). This model is a collective call for ESD to become part of school life and teaching. Chemistry education's contribution to sustainability in this model takes different forms, one of which is to influence practices and behavior, for example related to waste production, treatment and disposal, food, water and energy consumption, recycling mechanisms, etc. Burmeister et al. (2012) express that with this approach/model, "chemistry education no longer needs to stop at the point where teaching is limited to describing the chemical theories and knowledge behind sustainability issues and potential avenues of action" (p. 65), but transcends into an action-based pattern of living and learning. This understanding is similar to how Blum et al. (2017) conceptualize sustainable chemistry as a holistic approach that considers stakeholders and their responsibilities along the life cycle chain of chemicals, ensuring that conditions relating to health, environmental, social, research, science and economics, are considered and balanced within the capacity-limits of the planet. Overall, the central idea behind Burmeister et al.'s (2012) model 4 bears more transformational potential if achieved, though the chemistry teacher will find it difficult to lead such a transformational effort on his/her own. The enculturation connotations and demands that the model bears make it a huge task for the chemistry teacher. Nevertheless, the teacher could introduce such transformational behavior and practices in his/her chemistry class and expand into learning chemistry outside classroom environments and initiating chemistry mini-projects and inviting other students from other classes to participate and collaborating with those teachers with similar interests. Gradually, the individual effort could become a collective effort where everyone starts to recognize their role in contributing towards sustainability efforts.

10.3 Integration of Sustainability Issues in Chemistry Using a Systems Thinking Approach

To understand and be able to apply a systems thinking approach in chemistry teaching, we need to first understand what exactly is meant by "systems thinking". Several scholars have provided a definition for "Systems thinking" and Arnold and Wade (2015), having carried out a synthesis of several of these existing definitions or descriptions for what systems thinking is, proposed a definition that captures most of the elements considered by most other authors/scholars. It is this Arnold and Wade's (2015) definition that we consider here and use in our discussion about systems thinking and its application in chemistry teaching. Arnold and Wade (2015) conceptualize systems thinking as *"a system of thinking about systems"*. They describe systems thinking as "a set of synergistic analytic skills used to improve the capability of identifying and understanding systems, predicting their behaviors, and devising modifications to them in order to produce desired effects." (p. 675). The authors go further to provide a description for each of the different key terms/words in their definition. For example, they consider ***systems*** as groups or combinations of interrelated, interdependent, or interacting elements forming collective entities, ***synergistic***—the interaction of elements in a way that, when combined, produce a total effect that is greater than the sum of the individual elements, ***analytic skills***—skills that provide the ability to visualize, articulate, and solve both complex and uncomplicated problems and concepts, and make decisions that are sensible and based on available information. Such skills include demonstration of the ability to apply logical thinking to gathering and analyzing information, designing and testing solutions to problems, and formulating plans, ***identify***—to recognize as being a particular thing, ***understand***—to be thoroughly familiar with; apprehend clearly the character, nature, or subtleties of, ***Predict***—to foretell as a deducible consequence, and ***devise modifications***—to contrive, plan, or elaborate changes or adjustments (p. 675).

Applying the above definition and understandings to chemistry and chemistry education, chemistry as a field is a complete and dynamic system comprising several interconnected components (sub-systems). These components are coherently organized to form a comprehensive body of chemical knowledge with defined core practices. The chemistry system and its components interact with other systems (environmental, biological, physical, societal, etc.) and the inherent components of those. The effects of such continuous interactions between the components of the chemistry system and other systems are both positive and negative, beneficial and harmful (Mahaffy et al., 2018). Systems thinking as a chemistry teaching framework therefore is premised on the idea that the learning of chemistry should consider the central ideas and core practices in chemistry (science) and the interactions between these and specific components of other systems, explore the interactions, the positive and negative effects of existing interactions, and then draw on emerging understandings to inform and shape current and future societal practices and desired future interactions between chemistry and other systems. It is premised on the recognition that all elements within our planet—humans, animal, biological and environmental systems are intimately connected systems, and that exploring and understanding these existing interactions is needed if chemistry is to contribute effectively to the needed solutions to contemporary challenges (Orgill et al., 2019). That is, a systems thinking approach to chemistry teaching allows cross cutting concepts to be integrated to serve as a basis for developing systems thinking skills with which the individual can apply the central chemistry ideas and chemistry core practices to analyze, understand, and evaluate solutions to some of the existing sustainability problems and challenges (Talanquer & Szozda, 2024).

A framework for teaching chemistry using systems thinking is proposed as one of the outcomes of the IUPAC project: Systems Thinking in Chemistry for Sustainability: Toward 2030 and beyond (STCS 2030+) (Talanquer & Szozda, 2024). It is suggested and advanced as a guide into planning, implementation, and assessment of chemistry units, modules or lessons using a systems thinking approach. Moreover, the project team has developed exemplary educational resources to accompany the proposed framework, resources that chemistry teachers can use as examples when developing their own teaching plans that integrate and use systems thinking. These can be found by visiting their website at https://sastice.com/ and more information about the IUPAC project referred to above can be found at https://iupac.org/project/2020-014-3-050/. Note that these websites may or may not be active by the time you want to access them. Nevertheless, the IUPAC is a well-established framework organisation for the field of chemistry and hence these resources will be available if they are deemed and continue to be relevant to the chemistry education community. Moreover, details about the proposed framework and how it can be operationalized are described in a published article by Talanquer and Szozda (2024).

The framework proposed by The IUPAC—STCS 2030+ project for teaching chemistry using a systems thinking approach, comprises three iterative phases, planning, implementation and the assessment phase (Fig. 10.1). Four fundamental learning outcomes are considered by the authors (members of the project), which describe and represent the key knowledge and skills elements students should acquire, which once attained, will empower students with the ability to analyze, understand, and tackle global issues. These learning outcomes are (i) meaningful understanding of central ideas in chemistry, (ii) knowing how to apply science and engineering practices to investigate phenomena and design solutions, (iii) developing systems thinking skills necessary to analyze systems of interest, and (iv) acting

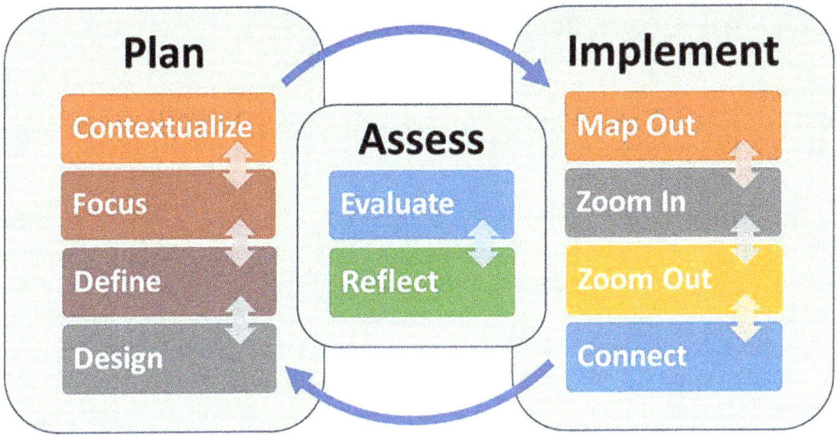

Fig. 10.1 Educational framework for teaching chemistry using a systems thinking approach (Talanquer & Szozda, 2024)

with socio-environmental responsibility (Talanquer & Szozda, 2024). These four fundamental learning outcomes that constitute the different elements that go into planning, implementing and assessment phases of the teaching and are taken up at different iterative phases of their proposed framework shown in Fig. 10.1.

Talanquer and Szozda (2024) provide a detailed description of the different elements of the framework with concrete steps about how a teacher goes about planning, implementing and through the assessment phase. Moreover, the authors also provide a concrete example of a complete teaching plan using the framework. This is in addition to other educational resources that I already mentioned that can be found at https://sastice.com/. Repeating this information here will be doing a disservice to the team of scholars who have already done such substantial work in preparing and making these resources available for chemistry teachers who wish to use the framework in their chemistry teaching. I would thus recommend reading the article "An Educational framework for teaching chemistry using a systems thinking approach" by Talanquer and Szozda (2024) for details about how this framework can be operationalized.

Moreover, while the above framework (Fig. 10.1) by The IUPAC—STCS 2030+ project seems to be the most well-structured, organized and underpinned by several educational perspectives, pedagogical and didactical understandings, it is not the only way that has been used to invoke and employ systems thinking in chemistry education. Some educators have approached and used systems thinking in different ways. These different ways could also be equally explored and used by teachers who wish to use systems thinking differently other than using the suggested framework in Fig. 10.1. Examples here include use of graphical tools such as systems oriented concept map extensions, causal loop diagrams, and stock and flow diagrams (Aubrecht et al., 2019), using system maps (Eaton et al., 2019), and the ChEMIST Table tool for designing or modifying instruction for a systems thinking (York & Orgill, 2020).

10.4 The Elliptic Model for Applying ESD in Chemistry Teaching

Jegstad and Sinnes (2015) modified and expanded the four basic models or strategies which Burmeister et al. (2012) suggested for integrating ESD in chemistry teaching (see earlier section), and suggested what they called the elliptic model for ESD in chemistry education. The model (Fig. 10.2) comprises five ellipses representing five different ESD categories, (i) chemical content knowledge, (ii) chemistry in context, (iii) chemistry's distinctiveness and methodological character, (iv) ESD competences and (v) Lived ESD. The conceptual understanding that underly the proposed model is that chemistry teaching regardless of the topic of choice, should promote development of respect and responsibility as well as those competences leaners need in their day-to-day living and as future citizens in a sustainable world (Jegstad & Sinnes, 2015). With this understanding, the authors share the view of other scholars who have emphasized a chemistry teaching that considers development of general skills (educational skills) (e.g.:, Holbrook, 2005) or graduate attributes (e.g.; Kayima, 2022), along with chemistry-specific skills and competencies.

The ***chemical content knowledge*** concerns chemistry issues that are relevant for SD. These for example could be water and energy resources, the effects of acid rain, greenhouse gases, the greenhouse gas effect, ozone and ozone layer depletion, oil exploration, CO_2 capture possibilities, renewable and nonrenewable resources, the relationship between abundance of natural minerals/chemical elements, their exploration and use in advancing green technologies, etc. ***Chemistry in context*** is about aligning chemistry concepts with relevant contexts to ensure the full understanding of sustainability issues under consideration. Chemistry teaching should be contextualized to provide for understanding the interactions between chemistry as a system with other systems such as society, and the environment. ***Chemistry's distinctiveness and methodological character*** entails applying sustainable practices in chemistry and addressing the nature of chemistry. With sustainable practices the authors point to teachers working within the framework principles of green chemistry and sustainable chemistry that we have already discussed, while nature of chemistry captures the nature and practices of science, since solutions towards sustainability practices are impacted directly by practices in chemistry and science generally (for example production processes, production of and treatment of chemical waste, etc.). The fourth ellipsis is about developing ***ESD competences***. Examples of these competences include ethical, communication and collaboration competences, decision-making skills, critical thinking and analysis skills, systems thinking competences, problem solving, creativity, action competence, normative competences, to mention but a few. The fifth ellipsis ***(Lived ESD)*** concerns ESD principles realized in classrooms and school cultures. They are lived experiences

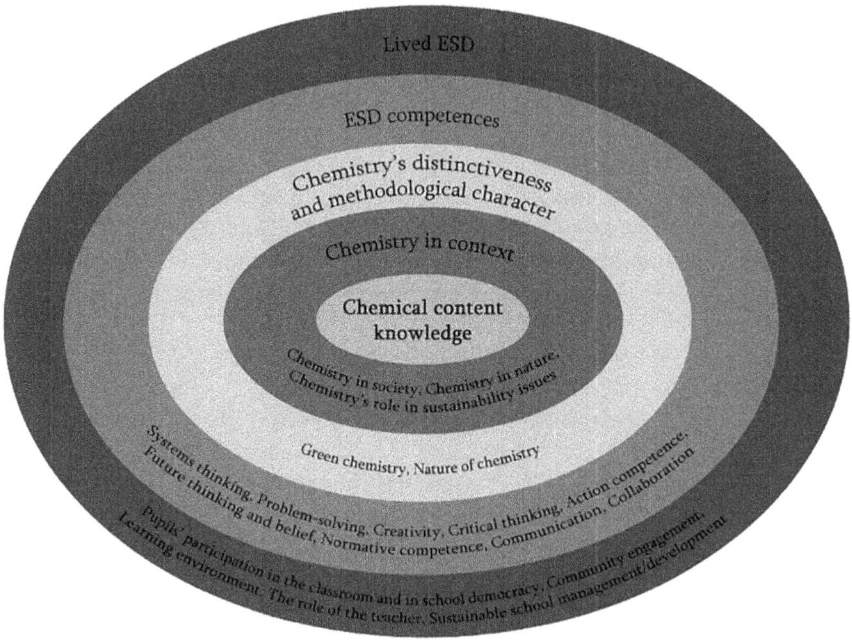

Fig. 10.2 The elliptic model for ESD in chemistry education: A model for planning and analyzing chemistry ESD by Jegstad and Sinnes (2015)

that align with the envisioned sustainable living or practices (Jegstad & Sinnes, 2015).

Our interest as teachers concerns how the above suggested model is operationalized in real teaching contexts. How can we apply or use the above model to plan and implement teaching in chemistry classrooms? The authors and developers of the model (Fig. 10.2) attempted to deal with this question of interest in rather an open way, leaving the huge responsibility on the side of the teacher. They express that, starting at the center of the model the teacher selects the chemistry topic as guided by the chemistry curriculum being used. The selected topic may or may not be directly associated with a particular sustainability issue. Regardless, the teacher should have as one of his/her goals to develop ESD skills along with chemistry knowledge and skills. The teacher, having selected the topic, goes on to select the context within which the topic will be situated. The teacher will then plan the teaching in a way that considers and follows green chemistry and sustainable chemistry principles and ideals. The authors express that ESD competences are taken up in the whole of the teaching process, in the used teaching and learning methods. The teacher should aim to develop and reinforce these ESD skills as part of the whole teaching and learning. The teacher ensures a lived ESD through providing a friendly and safe learning environment, starting within his/her own classroom and extending and promoting the practice throughout the whole school.

As mentioned earlier, the proposal by the authors on how to apply or use the model is very open and, if not, far from being concrete enough for a teacher to adopt it and use it in teaching. The outline they provide however gives some good insights to a teacher about a methodological path or roadmap to follow when planning to integrate ESD in a chemistry teaching lesson, and what important aspects to consider such as working within the principles of green and sustainable chemistry, contextualizing the selected chemistry topics, linking the selected chemistry topic with the relevant sustainability issues, and ensuring the development of ESD skills along with subject canonical knowledge. It is therefore up to the teacher to decide how all these elements suggested and included in the model could and will be realized.

10.5 The Project- and Problem-Based Learning Model

Both project-based learning (PjBL) and problem-based learning (PBL) are inquiry-based learning models where students work (in teams or groups) to explore and find solutions to designated problems while the teacher takes on the role of a facilitator. Both are student-centered pedagogies aimed to foster students' deeper learning through exploring real-world problems (Nagarajan & Overton, 2019). With PBL students are provided with a problem that relates to real-life situations and they work towards finding a solution to the problem by applying their knowledge (Vergara-Castañeda et al., 2021). The problems teachers provide are often open-ended and are contextualized in real-life scenarios (Schmidt, 2012). In working to find solutions to these open-ended problems, students actualize their knowledge both canonical and experiential knowledge and thereby establishing a link between theory and practice throughout the process. As such, PBL provides for students to actively participate, train to be autonomous, creative, collaborative, and to take responsibility (Tuan et al., 2020). Moreover, as students work and accomplish tasks, they get to improve their critical thinking, problem solving, and acquire soft skills (Aidoo et al., 2016; Vergara-Castañeda et al., 2021). With PjBL, the project or designated problem is more rigorous, extended and often more complex.

While PjBL and PBL share some common features to the extent that project-based and problem-based are used interchangeably by some educators (McLaughlin et al., 2024), the two pedagogies have some key elements that distinguish them from one another. With PjBL for example, the project can be designed to be accomplished over an extended period with the goal of students arriving at a specific product or developing a complete presentation at the end of the project. It is often structured around a specific theme or a question that is contextualized in real-life, and students are expected to employ scientific practices and principles as they work towards the solution. The process involves for example students engaging in research, problem solving, collaborative work and critical thinking and reflection, and they are expected to use or apply knowledge and experiences from different disciplines. Nagarajan and Overton (2019), having reviewed several literature studies on the use of PjBL, summarize five principles of a PjBL pedagogy, namely, (i) a clear open-ended and realistic question (s) that shapes and drives students' activities (the driving question or challenge), (ii) using scientific practices—the process is and should be driven by scientific thinking and operational thinking—practices such as setting the hypothesis, identifying variables, brainstorming, engaging in problem-solving, etc., (iii) a sustained inquiry process from start to end, involving assessment and feedback, (iv) use of relevant technologies to support the inquiry process, and (v) deriving at and presenting the final product. PBL on the other hand does not go as far as invoking or demanding an account of all the above. The activities with the PBL pedagogy are conceived as rather short duration activities centered around a specific problem and the outcome may not be predefined. Students often work towards finding the solution to the specified problem. In so doing, they engage in research, draw on their knowledge and experiences, use available resources, and as they do this, they acquire and enhance their thinking, analysis and collaboration skills.

Whether we consider the distinction between PjBL and PBL in planning to integrate sustainability issues in our chemistry teaching will depend mainly on how far we are willing to engage our students into such a student-centered expedition. Both PjBL and PBL pedagogies allow teachers to connect chemistry to real world problems and to provide students with the opportunity to explore, engage with, and solve complex problems related to sustainability. This leads to the question of interest, which concerns how chemistry teachers can use PBL/PjBL to plan a chemistry teaching that addresses and equips students with both chemistry knowledge and skills as well as sustainable knowledge and skills. Again, here the teacher's creativity, knowledge and the ability to design either a project or develop a relevant problem and structuring the different parts, stages and expectations of whole activity/project is needed. The underlying principles are the same but there are different ways to organize, implement and achieve target learning goals using PjBL/PBL pedagogies. We can draw on some examples from literature for insight into how to get started.

Overton and Randles (2015) for example describe six simple steps involved in a PBL pedagogy, (i) students are divided into groups and presented with a problem, (ii) students brainstorm to clarify and understand the nature and scope of the problem and identify their learning needs, (iii) students delegate roles and responsibilities within their respective groups and share knowledge and experiences aligned with the problem they seek to solve, (iv) outside the classroom, students engage in independent study, (v) they come together in their respective groups to share and critically evaluate the resources and information they each gathered, and (vi) they work towards a solution (p.252). Accordingly, the process involves rounds of independent and group explorations until a solution is collectively derived at. The teacher provides support and guides students along the way as deemed necessary. Molina-Torres (2022) describes in her work, four characteristics of PjBL framework, namely (i) collaborative learning, (ii) competency-based and interdisciplinary work, (iii) group and individual decision-making, and (iv) real-world problem-solving. These four characteristics are well reflected in the six steps described by Overton and Randles (2015), and inform on the target goal and potential of a PjBL pedagogy to develop canonical knowledge, sustainability and soft skills.

Also, Tuan et al. (2020) and several other scholars describe PjBL to involve at least five phases. First is a project construction phase that involves the instructor developing and presenting the project problem to the students. The expectation is that the presented problem should be characteristic of a real -world problem, anchored in a chemistry concept or theme of interest and should target one or more sustainability issues. The second phase is planning the implementation. According to Tuan et al. (2020), this phase involves students discussing in their groups (it is assumed that these groups are already in place), study the given problem and devise and discuss possible ways to solve the problem. This phase combines Overton and Randle's (2015) second steps ii and iii, as it also involves students sharing and delegating responsibilities amongst themselves within their respective groups. Phase three is the project implementation phase. This is when students implement and carry out the activities as agreed upon in their respective groups. Tuan et al. (2020) stress that students collaborate within and across the groups during the implementation phase, while Overton and Randles (2015) talks about and stress the importance of independent study (step iv) and collaborative study (step v) as main components of the implementation phase. Phase four involves reporting project results. There are different ways to report project results. The teacher can guide students into the most preferred or suitable forms with respect to the goal of the project or its target learning outcomes. The fifth and last phase is the evaluation of project results. The assessment can take different forms. For example, individual students can be asked to give feedback about the whole project, its goals and results, the individual groups could be asked to do the same, etc. Accordingly, evaluation or assessment should be geared toward the target learning goals of the project. This is where assessment of learning should come to light. Indeed, for a competent and experienced instructor, assessment should not be only at the end but should be considered and baked into all phases to track and support students in training, acquiring, and developing the target knowledge and skills. In other words, both assessment of learning and assessment for learning should be considered.

To summarize, a central feature of the PjBL/PBL pedagogies is that it starts with a teacher being able to come up with a relevant problem that will provide students with opportunities to learn chemistry, apply aspects of chemistry in real-life contexts, apply green and sustainable chemistry principles and ideals, train in and acquire sustainability skills. As from the examples above, the steps or phases described to constitute the PjBL/PBL pedagogies are simple, iterative, open and non-restrictive. Note that these are simply described stages of how you can structure the activities, which activity comes first and what should follow, etc. They are thus simply guidlines that you can use as a teacher and the duty of developing the contents to fit into or serve these guidelines is yours as a teacher. This implies that you must or should seek to learn and be able plan activities yourself using the proposed phases/steps as a guide. There are a range of real-life topics to connect with specific chemistry aspects/concepts/topics. These could for example include energy consumption, food, waste management, environmental impact, ecology, health issues, etc. Moreover, there are several other sustainable development themes of interest to chemistry and chemistry education, which can be used to plan a PjBL/PBL session. It should be noted also that the examples provided above are not the only methodological roadmaps to carrying out a PjBL/PBL session. There is a quite a range of other ways exemplified in educational literature with which one could implement PjBL/PBL pedagogies, which can be drawn on to plan and carryout the intended teaching (see for example, Günter et al., 2017; Paristiowati et al., 2022; Salgado-Chavarría & Palacios-Alquisira, 2021; Vergara-Castañeda et al., 2021).

10.6 Action-Competence Teaching Approach—Action-Oriented Knowledge on Sustainability Issues

An action-competence approach is driven by a desire towards a teaching and learning that is more action-oriented (Jensen & Schnack, 1997). Action-competence (AC) involves the ability to, willingness to, and being qualified to act (Jensen & Schnack, 1997). That is, the readiness to act in ways that meet challenges in specific situations (Blomhøj & Jensen, 2003; Van Poeck et al., 2024). AC "describes peoples' ability to act towards solving controversial problems, when it is necessary to take decisions based on critical thinking and incomplete knowledge" (Lampert et al., 2023, p. 3). Accordingly an action competent person "is committed and passionate about solving a societal issue, has the relevant knowledge about the issue at stake as well as about the democratic processes involved, takes a critical but positive stance toward different ways for solving it, and has confidence in their own skills and capacities for changing the conditions for the better."(Sass et al., 2020, p. 300). Torsdottir et al. (2024) conceptualize AC as a competence comprising three sub-constructs, (i) knowledge of action possibilities, (ii) confidence in one's own influence, and (iii) willingness to act.

Several researchers and educators strongly express that it is not sufficient for learners to only learn about sustainable development, but they should experience taking part in concrete and authentic actions relating to sustainability (Sinakou et al., 2019; Torsdottir et al., 2024). Jensen (2002, 2004) puts across four dimensions (questions) that should be considered in any characteristically action-oriented teaching that considers integrating sustainability issues. These include, (i) what kind of problem is it?—knowledge about the problem and its effects, (ii) why do we have the problem we have?—knowledge about the root causes, (iii) how do we change things?—knowledge about strategies for change, and (iv) where do we want to go?—knowledge about alternatives and visions (Jensen, 2002, pp. 229–231; Van Poeck et al., 2024, p. 335). These four dimensions inform and underly action-oriented teaching. They are dimensions that illustrate different perspectives on the kinds of knowledge with which environmental (or more broadly sustainability issues) can be viewed and analyzed.

Moreover, the rationale towards action-oriented teaching by educators is that the way sustainable issues are taken up and integrated in traditional (current) teaching tends to concentrate mainly on students attaining or acquiring knowledge about the existing sustainability issues/problems and how these issues/problems are evolving resulting from human actions, etc. Students are provided with "a sound and well-founded understanding of, and insight into, the effects of environmental problems on individuals and societies, and the natural world" (Jensen, 2004, p. 406) and insights into the extent the environmental conditions are worsening. The teaching does not however extend to consider and address all the mentioned four dimensions, which accordingly would provide learners the readiness needed to act in ways that meet the different challenges and the ability to contribute to shaping a sustainable future (Jensen, 2002, 2004; Van Poeck et al., 2024). Because the teaching focuses more on knowledge about the existence of problems and their extent, and less on training in competencies needed to take action, worry and action-paralysis develops among learners (Jensen, 2004). The strong argument therefore is that, any teaching aimed at addressing environmental or altogether sustainability issues, should develop students' abilities to take action themselves, that is, their *action-competence.*

Undoubtedly learners need action competences to be able to act and take part in important decisions and sustainability related actions, and for themselves to live and act in ways that promote sustainable development efforts. The question that must be answered then concerns how chemistry teachers can plan chemistry teaching that supports students to develop the needed action competences. As earlier mentioned, a participatory approach (Wals & Mathie, 2020), that gives students the opportunity to experience taking part in concrete and authentic actions, can provide learning experiences that trigger the three subconstructs of AC described by Torsdottir et al. (2024). That is, through taking part in concrete and authentic actions, students are expected to acquire not only canonical knowledge but also knowledge about sustainability issues and challenges and the possible actions and individual obligations, to reinforce their own influence and confidence, and their willingness to do their part as society members.

Three teaching approaches for ESD teaching and learning, are advanced in educational literature concerned with ESD, namely a holistic (holism), a pluralistic, and a transformative or transactional approach (e.g.:, Borg et al., 2014; Boström et al., 2018; Hopkins, 2012; Olsson et al., 2022; Pauw et al., 2015; Sinakou et al., 2019; Öhman & Östman, 2019; Östman et al., 2019). Holism in ESD represents a teaching approach integrating multiple perspectives—environmental, economic and social, and their interactions across time and space. This approach offers the opportunity for students to understand that sustainability issues often arise from conflicts between different dimensions. It is an interdisciplinary approach that attempts to link local and global issues and their implications (that's the past, present and future implications), In so doing, a comprehensive understanding of sustainability problems is offered and thereby allowing students to visualize how various systems and interests are interconnected (Borg et al., 2014; Hopkins, 2012; Pauw et al., 2015). A pluralistic approach on the other hand is a collective and participatory teaching approach "where students are involved in discussions where different views and values are acknowledged in relation to the sustainability issue at hand." (Olsson et al., 2022, p. 407). Pluralism in ESD focuses on supporting and developing competences that enable students to make well-informed decisions in relation to sustainability issues (Sinakou et al., 2019). A transformative/transactional teaching approach emphasizes a form of teaching that changes the way students think, feel and act regarding sustainability. The approach "includes components such as critical thinking, a participatory (student-centered) approach, collective learning, multi-method approaches, decision-making, and/or value-based learning" (Boström et al., 2018; Olsson et al., 2022, p. 407).

While the above three teaching approaches seem feasible, achieving such instructional organizations by chemistry teachers faced with different curricula demands cannot come without challenges. For example, the holistic integration of multiple perspectives—environmental, economic and social, and their interactions across time and space requires teachers to understand what this means in the chemistry teaching context and the amount of scaling down that is required. The alternative to this kind of effort to achieve the interdisciplinary nature of actions in sustainability is to embrace and use a systems thinking approach that was discussed earlier. Moreover, the systems thinking framework suggested by The IUPAC—STCS 2030+ project (Fig. 10.1) (Talanquer & Szozda, 2024), allows students to visualize how various systems and interests are interconnected, and it provides a clearer guide to teaching both chemistry concepts and sustainability issues. The pluralistic approach is also complex and time consuming for teachers to achieve this kind of teaching. It is a collective and participatory teaching approach; students are involved in discussions involving different views and values relating to sustainability issues at hand. With this formulation, it is easy to think that this is easy to realize in practice, however, it again goes down to teachers being able to scale down and focus the directions of interest, along with the kind of actions or behaviors that they want their students to acquire. Our goal is to integrate sustainability issues into chemistry. This means that we consider both chemistry knowledge and sustainability issues in our teaching. The pluralistic approach in chemistry hence requires that we draw on existing pedagogies such as situational learning and context-based learning, to enable us to capture both the chemistry that we want our students to learn and those sustainability issues of interest.

A transformative/transactional approach is about changing the way students think, feel and act regarding sustainability. It is a student-centered approach that opens for multi-methods to be employed. Understood this way, the approach is feasible and can be used in chemistry teaching without much difficulty. However, the approach relies on high levels of student engagement and teacher support. Each one of the approaches we have discussed before in this chapter can be employed or used by the teachers with the goal of changing how students think, feel or act. It only calls for proper planning and having well-organized and structured activities for students. Note that a transformative/transactional approach emphasizes deep internal change, which is abstract and difficult to measure or evaluate. However, engaging learners in appropriate learning experiences is poised to transform their thinking and actions, be it in the short or long run.

Finally, as mentioned earlier, our interest is not only learning about and understanding sustainability issues but integrating sustainability issues into our teaching and learning of chemistry. It means addressing canonical subject knowledge and skills as well as sustainability knowledge, skills and other attributes. We thus need to employ those teaching methods and techniques that will enable us to achieve this goal. What we achieve from the above conceptualizations regarding action competence in ESD teaching is that it is not sufficient to only learn about sustainability issues, but students should be enabled with knowledge and relevant skills (competences) to enable them to know what to do in response to sustainability issues, how to act, and to reinforce their willingness to act. This means that we can use any student-centered approach, starting with those that have been discussed in this book, to derive at our teaching goals if we can justify our teaching decisions within well-established pedagogical frameworks. Moreover, it is possible to develop a teaching plan based on the four questions (dimensions) by Jensen (Jensen, 2002, pp. 229–231; Van Poeck et al., 2024, p. 335), namely (i) what kind of kind of problem is it? (ii) why do we have the problem we have? (iii) how do we change things? and (iv) where do we want to go? However, this requires that we expand the meanings associated with each dimension if we are to achieve a teaching that considers both chemistry knowledge and skills and sustainability issues.

Let us take for example "Plastics and humans" as our overall theme and work towards a chemistry lesson plan or module that integrates both chemistry and sustainability issues. We start by listing the different aspects that will together form the structure of the lesson or module. These should include,

1. Deciding the chemistry content—the topic in chemistry to connect with plastics—here it is easy to think about and consider the chemistry behind plastics—polymers and polymerization, chemical composition and properties of these compounds, etc. Notice that the choice of the topic is informed by the governing chemistry curriculum.
2. Decide on the sustainability content—the focus is plastics but what exactly is to be addressed about plastics—consider for example plastic and the threat to the environment, different forms of plastics, structural composition, durability, biodegradability, etc.
3. Decide on the source for firsthand experiences—the real-life context—connecting plastics and students' experiences. Here the resources factor plays an important role. You must assess what resources are at your disposal as all planning is dependent on what is available, accessible and not.
4. Define learning goals/aims—those targeting chemistry knowledge and skills and those targeting sustainability knowledge, skills and competencies

5. Plan and develop learning activities—according to the defined learning goals. The lesson can be divided into different parts—that's the theory part, the practical part and the extended project part
6. Develop an assessment strategy and assessment activities or tools, targeting both assessment for learning and assessment of learning as deemed necessary. The assessment should target both chemistry related knowledge and skills and sustainability related knowledge, skills and competencies

Considering the list above, you realize that the lesson should be comprehensive to cater for both chemistry-learning interest and sustainability learning interests. In fact, it leads us to the idea that an action-competence approach illuminates what should be the end goal of learning and that the four dimensions suggested by Jensen (2002, 2004), should inform both the planning and the course of implementation. Thus, a combination of didactical approaches is needed. This is visible from the drafted list above, where a lesson according to the above list can have three parts—(i) the in-class teaching session (where students learn about and go through the chemistry behind plastics, etc., (ii) the out-of-class exploration/in-class experimental part (where students explore plastics in nature, test for presence of micro-plastics in water sources, etc., and (iii) a project-based learning component-where students for example work on themes such as– efficient plastic waste reduction mechanisms within community Q. Moreover, it is possible to extend and plan more than the three suggested module parts. Note also that there are several (possible) different target learning goals and hence it is these well-defined target learning goals that will shape your planning and the total number of different modules to develop. The work is thus complex and demands both competence and a strong willingness (motivation) on the side of the teacher.

References

Aidoo, B., Boateng, S. K., Kissi, P. S., & Ofori, I. (2016). Effect of problem-based learning on students' achievement in chemistry. *Journal of Education and Practice, 7*(33), 103–108.

Anastas, P. T. (2003). *Meeting the challenges to sustainability through green chemistry.* Green Chemistry.

Anastas, P. T., & Warner, J. C. (1998). *Green chemistry: Theory and practice.* Oxford University Press.

Anastas, P. T., & Zimmerman, J. B. (2018). The United Nations sustainability goals: How can sustainable chemistry contribute? *Current Opinion in Green and Sustainable Chemistry, 13,* 150–153. https://doi.org/10.1016/j.cogsc.2018.04.017

Aoki, E., Rastede, E., & Gupta, A. (2022). Teaching sustainability and environmental justice in undergraduate chemistry courses. *Journal of Chemical Education, 99*(1), 283–290. https://doi.org/10.1021/acs.jchemed.1c00412

Armstrong, L. B., Rivas, M. C., Zhou, Z., Douskey, M. C., & Baranger, A. M. (2024). Behind the scenes of teaching green: An Iterative approach to curriculum design and implementation in the general chemistry laboratory. *Journal of Chemical Education, 101*(8), 3264–3275. https://doi.org/10.1021/acs.jchemed.4c00176

Arnold, R. D., & Wade, J. P. (2015). A Definition of systems thinking: A systems approach. *Procedia Computer Science, 44,* 669–678. https://doi.org/10.1016/j.procs.2015.03.050

Aubrecht, K. B., Dori, Y. J., Holme, T. A., Lavi, R., Matlin, S. A., Orgill, M., & Skaza-Acosta, H. (2019). Graphical tools for conceptualizing systems thinking in chemistry education. *Journal of Chemical Education, 96*(12), 2888–2900. https://doi.org/10.1021/acs.jchemed.9b00314

Azizah, U., & Yonata, B. (2023). The Integration of green chemistry principles in basic chemistry learning to support achievement of Sustainable Development Goals (SDGs) through education. *Journal of technology and science education, 13*(1), 233–254.

Blomhøj, M., & Jensen, T. H. (2003). Developing mathematical modelling competence: Conceptual clarification and educational planning. *Teaching Mathematics and its Applications: An International Journal of the IMA, 22*(3), 123–139. https://doi.org/10.1093/teamat/22.3.123

Blum, C., Bunke, D., Hungsberg, M., Roelofs, E., Joas, A., Joas, R., Blepp, M., & Stolzenberg, H.-C. (2017). The concept of sustainable chemistry: Key drivers for the transition towards sustainable development. *Sustainable Chemistry and Pharmacy, 5*, 94–104. https://doi.org/10.1016/j.scp.2017.01.001

Borg, C., N., G., O., H. H., & Bergman, E. (2014). Subject- and experience-bound differences in teachers' conceptual understanding of sustainable development. *Environmental Education Research, 20*(4), 526-551. https://doi.org/10.1080/13504622.2013.833584

Boström, M., Andersson, E., Berg, M., Gustafsson, K., Gustavsson, E., Hysing, E., Lidskog, R., Löfmarck, E., Ojala, M., Olsson, J., Singleton, B. E., Svenberg, S., Uggla, Y., & Öhman, J. (2018). Conditions for transformative learning for sustainable development: A theoretical review and approach. *Sustainability, 10*(12), 4479. https://www.mdpi.com/2071-1050/10/12/4479

Burmeister, M., Rauch, F., & Eilks, I. (2012). Education for Sustainable Development (ESD) and chemistry education [10.1039/C1RP90060A]. *Chemistry Education Research and Practice, 13*(2), 59–68. https://doi.org/10.1039/C1RP90060A

Cáceres-Jensen, L., Rodríguez-Becerra, J., Jorquera-Moreno, B., Escudey, M., Druker-Ibañez, S., Hernández-Ramos, J., Díaz-Arce, T., Pernaa, J., & Aksela, M. (2021). Learning reaction kinetics through sustainable chemistry of herbicides: A case study of preservice chemistry teachers' perceptions of problem-based technology enhanced learning. *Journal of Chemical Education, 98*(5), 1571–1582. https://doi.org/10.1021/acs.jchemed.0c00557

Eaton, A. C., Delaney, S., & Schultz, M. (2019). Situating sustainable development within secondary chemistry education via systems thinking: A depth study approach. *Journal of Chemical Education, 96*(12), 2968–2974. https://doi.org/10.1021/acs.jchemed.9b00266

Eilks, I., & Linkwitz, M. (2022). Greening the chemistry curriculum as a contribution to education for sustainable development: When and how to start? *Current Opinion in Green and Sustainable Chemistry, 37*, 100662. https://doi.org/10.1016/j.cogsc.2022.100662

Günter, T., Akkuzu, N., & Alpat, Ş. (2017). Understanding 'green chemistry' and 'sustainability': an example of problem-based learning (PBL)*. *Research in Science & Technological Education, 35*(4), 500–520. https://doi.org/10.1080/02635143.2017.1353964

Holbrook, J. (2005). Making chemistry teaching relevant. *Chemical education international, 6*(1), 1–12.

Hopkins, C. (2012). Reflections on 20+ Years of ESD. *Journal of Education for Sustainable Development, 6*(1), 21–35. https://doi.org/10.1177/097340821100600108

Horváth, I. T. (2018). Introduction: Sustainable chemistry. *Chemical Reviews, 118*(2), 369–371. https://doi.org/10.1021/acs.chemrev.7b00721

Idul, J. J. A., Jaculbe, Q. M. I., Lucine, N. S., Canama, M. D., Galve, M. D., Sayson, C. L., Suico, M. P., Santiago, K. Z. D., & Walag, A. M. P. (2025). Green modules: Integrating green and sustainable chemistry principles to secondary chemistry modules through process-oriented guided inquiry learning. *Journal of Chemical Education, 102*(3), 1104–1116. https://doi.org/10.1021/acs.jchemed.4c01360

Idul, J. J. A., & Walag, A. M. P. (2024). Integrating Green chemistry and sustainability principles to a secondary science curriculum: A mixed-methods needs assessment. *Journal of Chemical Education, 101*(7), 2765–2778. https://doi.org/10.1021/acs.jchemed.4c00341

Iyere, P. A. (2008). Chemistry in sustainable development and global environment. *Journal of Chemical Education, 85*(12), 1604. https://doi.org/10.1021/ed085p1604

Jegstad, K. M., & Sinnes, A. T. (2015). Chemistry teaching for the future: A model for secondary chemistry education for sustainable development. *International Journal of Science Education, 37*(4), 655–683. https://doi.org/10.1080/09500693.2014.1003988

Jensen, B. B. (2002). Knowledge, action and pro-environmental behaviour. *Environmental education research, 8*(3), 325–334. https://doi.org/10.1080/13504620220145474

Jensen, B. B. (2004). Environmental and health education viewed from an action-oriented perspective: A case from Denmark. *Journal of Curriculum Studies, 36*(4), 405–425. https://doi.org/10.1080/0022027032000167235

Jensen, B. B., & Schnack, K. (1997). The action competence approach in environmental education. *Environmental education research, 3*(2), 163–178. https://doi.org/10.1080/1350462970030205

Juntunen, M. K., & Aksela, M. K. (2014). Education for sustainable development in chemistry—Challenges, possibilities and pedagogical models in Finland and elsewhere [10.1039/C4RP00128A]. *Chemistry Education Research and Practice, 15*(4), 488–500. https://doi.org/10.1039/C4RP00128A

Kayima, F. (2022). Teaching graduate attributes along with subject content: Perspectives from science teacher educators. *Nordic Studies in Science Education, 18*(3), 290–304.

Lampert, P., Goulson, D., Olsson, D., Piccolo, J., & Gericke, N. (2023). Sustaining insect biodiversity through Action Competence —An educational framework for transformational change. *Biological Conservation, 283*, 110094. https://doi.org/10.1016/j.biocon.2023.110094

Lozano, R. (2008). Envisioning sustainability three-dimensionally. *Journal of Cleaner Production, 16*(17), 1838–1846. https://doi.org/10.1016/j.jclepro.2008.02.008

Mahaffy, P. G., Krief, A., Hopf, H., Mehta, G., & Matlin, S. A. (2018). Reorienting chemistry education through systems thinking. *Nature Reviews Chemistry, 2*(4), 0126. https://doi.org/10.1038/s41570-018-0126

Mahaffy, P. G., Matlin, S. A., Holme, T. A., & MacKellar, J. (2019). Systems thinking for education about the molecular basis of sustainability. *Nature Sustainability, 2*(5), 362–370. https://doi.org/10.1038/s41893-019-0285-3

Mahaffy, P. G., Matlin, S. A., Whalen, J. M., & Holme, T. A. (2019). Integrating the molecular basis of sustainability into general chemistry through systems thinking. *Journal of Chemical Education, 96*(12), 2730–2741. https://doi.org/10.1021/acs.jchemed.9b00390

McLaughlin, S., Amir, H., Garrido, N., Turnbull, C., Rouncefield-Swales, A., Swadźba-Kwaśny, M., & Morgan, K. (2024). Evaluating the impact of project-based learning in supporting students with the a-level chemistry curriculum in Northern Ireland. *Journal of Chemical Education, 101*(2), 537–546. https://doi.org/10.1021/acs.jchemed.3c01184

Molina-Torres, M.-P. (2022). Project-based learning for teacher training in primary education. *Education Sciences, 12*(10), 647. https://www.mdpi.com/2227-7102/12/10/647

Moog, R. S., & Spencer, J. N. (2008). POGIL: An overview. In R. S. Moog & J. N. Spencer (Eds.), *Process Oriented Guided Inquiry Learning (POGIL)* (Vol. 994, pp. 1–13). American Chemical Society. https://doi.org/10.1021/bk-2008-0994.ch001

Moog, R. S., Spencer, J. N., & Straumanis, A. R. (2006). Process-oriented guided inquiry learning: POGIL and the POGIL project. *Metropolitan Universities, 17*(4), 41–52.

Nagarajan, S., & Overton, T. (2019). Promoting systems thinking using project- and problem-based learning. *Journal of Chemical Education, 96*(12), 2901–2909. https://doi.org/10.1021/acs.jchemed.9b00358

Olsson, D., N., G., & Boeve-de Pauw, J. (2022). The effectiveness of education for sustainable development revisited—A longitudinal study on secondary students' action competence for sustainability. *Environmental Education Research, 28*(3), 405–429. https://doi.org/10.1080/13504622.2022.2033170

Orgill, M., York, S., & MacKellar, J. (2019). Introduction to systems thinking for the chemistry education community. *Journal of Chemical Education, 96*(12), 2720–2729. https://doi.org/10.1021/acs.jchemed.9b00169

Overton, T. L., & Randles, C. A. (2015). Beyond problem-based learning: using dynamic PBL in chemistry [10.1039/C4RP00248B]. *Chemistry Education Research and Practice, 16*(2), 251–259. https://doi.org/10.1039/C4RP00248B

Paristiowati, M., Rahmawati, Y., Fitriani, E., Satrio, J. A., & Putri Hasibuan, N. A. (2022). Developing preservice chemistry teachers' engagement with sustainability education through an online project-based learning summer course program. *Sustainability, 14*(3), 1783. https://www.mdpi.com/2071-1050/14/3/1783

Pauw, J. B.-D., Gericke, N., Olsson, D., & Berglund, T. (2015). The effectiveness of education for sustainable development. *Sustainability, 7*(11), 15693–15717. https://www.mdpi.com/2071-1050/7/11/15693

Rockström, J., Steffen, W., Noone, K., Persson, Å., Chapin, F. S., Lambin, E. F., Lenton, T. M., Scheffer, M., Folke, C., Schellnhuber, H. J., Nykvist, B., de Wit, C. A., Hughes, T., van der Leeuw, S., Rodhe, H., Sörlin, S., Snyder, P. K., Costanza, R., Svedin, U., … Foley, J. A. (2009). A safe operating space for humanity. *Nature, 461*(7263), 472–475. https://doi.org/10.1038/461472a

Salgado-Chavarría, D., & Palacios-Alquisira, J. (2021). Problem-based learning approach to review the principles of green chemistry applied to a polycondensation reaction. *Science Education International, 32*(2), 107–113.

Sass, W., Jelle, B.-D.P., Daniel, O., Niklas, G., Sven, D. M., & Van Petegem, P. (2020). Redefining action competence: The case of sustainable development. *The Journal of Environmental Education, 51*(4), 292–305. https://doi.org/10.1080/00958964.2020.1765132

Schmidt, H. G. (2012). A brief history of problem-based learning. In G. O'Grady, E. H. J. Yew, K. P. L. Goh, & H. G. Schmidt (Eds.), *One-day, one-problem: An approach to problem-based learning* (pp. 21–40). Springer Singapore. https://doi.org/10.1007/978-981-4021-75-3_2

Sinakou, E., Donche, V., Boeve-de Pauw, J., & Van Petegem, P. (2019). Designing powerful learning environments in education for sustainable development: A conceptual framework. *Sustainability, 11*(21), 5994. https://www.mdpi.com/2071-1050/11/21/5994

Steffen, W., Richardson, K., Rockström, J., Cornell, S. E., Fetzer, I., Bennett, E. M., Biggs, R., Carpenter, S. R., de Vries, W., de Wit, C. A., Folke, C., Gerten, D., Heinke, J., Mace, G. M., Persson, L. M., Ramanathan, V., Reyers, B., & Sörlin, S. (2015). Planetary boundaries: Guiding human development on a changing planet. *Science, 347*(6223), 1259855. https://doi.org/10.1126/science.1259855

Talanquer, V., & Szozda, A. R. (2024). An educational framework for teaching chemistry using a systems thinking approach. *Journal of Chemical Education, 101*(5), 1785–1792. https://doi.org/10.1021/acs.jchemed.4c00216

Timmer, B. J. J., Schaufelberger, F., Hammarberg, D., Franzén, J., Ramström, O., & Dinér, P. (2018). Simple and effective integration of green chemistry and sustainability education into an existing organic chemistry course. *Journal of Chemical Education, 95*(8), 1301–1306. https://doi.org/10.1021/acs.jchemed.7b00720

Torsdottir, A. E., Olsson, D., & Sinnes, A. T. (2024). Developing action competence for sustainability—Do school experiences in influencing society matter? *Global Environmental Change, 86*, 102840. https://doi.org/10.1016/j.gloenvcha.2024.102840

Tuan, N. N., Hanh, B. T., & Ninh, T. T. (2020). Project based learning in general chemistry to develop the problem-solving and creativity. *American Journal of Educational Research, 8*(7), 475–479. http://pubs.sciepub.com/education/8/7/4

Van Poeck, K., Vandenplas, E., & Östman, L. (2024). Teaching action-oriented knowledge on sustainability issues. *Environmental education research, 30*(3), 334–360. https://doi.org/10.1080/13504622.2023.2167939

Vergara-Castañeda, A., Chávez-Miyauchi, T.-E., Benítez-Rico, A., & Ogando-Justo, A.-B. (2021). Implementing project-based learning as an effective alternative approach for chemistry practical courses online. *Journal of Chemical Education, 98*(11), 3502–3508. https://doi.org/10.1021/acs.jchemed.1c00379

Wals, A. E. J., & Mathie, R. G. (2020). Whole school responses to climate urgency and related sustainability challenges. In M. A. Peters & R. Heraud (Eds.), *Encyclopedia of educational innovation* (pp. 1–8). Springer Singapore. https://doi.org/10.1007/978-981-13-2262-4_263-1

Wissinger, J. E., Visa, A., Saha, B. B., Matlin, S. A., Mahaffy, P. G., Kümmerer, K., & Cornell, S. (2021). Integrating sustainability into learning in chemistry. *Journal of Chemical Education, 98*(4), 1061–1063. https://doi.org/10.1021/acs.jchemed.1c00284

York, S., & Orgill, M. (2020). ChEMIST Table: A tool for designing or modifying instruction for a systems thinking approach in chemistry education. *Journal of Chemical Education, 97*(8), 2114–2129. https://doi.org/10.1021/acs.jchemed.0c00382

Zuin, V. G., Eilks, I., Elschami, M., & Kümmerer, K. (2021). Education in green chemistry and in sustainable chemistry: perspectives towards sustainability [10.1039/D0GC03313H]. *Green Chemistry, 23*(4), 1594–1608. https://doi.org/10.1039/D0GC03313H

Öhman, J., & Östman, L. (2019). Different teaching traditions in environmental and sustainability education. In V. P. Östman, K. L. & J. Öhman (Eds.), *Sustainable development teaching* (pp. 70–82). Routledge.

Östman, L., Van Poeck, K., & Öhman, J. (2019). Principles for sustainable development teaching. In K. V. Poeck, L. Östman, & J. Öhman (Eds.), *Sustainable development teaching* (pp. 40–55). Routledge.

School Practice: Learning in and from Practice

11

Abstract

School practice, also known as *teaching practicum* in some teacher educational contexts, is a formally organized and well-structured educational activity that teacher students are expected to successfully complete as part of their teacher professional training. It is purposely organized to provide teacher trainees with a basis and platform to practice teaching in a real context, namely the classroom. In so doing, trainees are expected to develop into effective practicing teachers, demonstrate the attained pedagogical and didactical competences, and to display their progress in developing as professional teachers. It is through school practice (practicum) that trainees get a chance to put their personal theories about teaching to test (Kayima, 2022), a process which triggers and results in transitioned thinking about the nature and practice of teaching. This chapter addresses the potential contributions of school practice towards a trainee's professional development into a competent teacher practitioner. It provides a background regarding the central learning elements of school practice as one component of teacher professional training and development, and how teacher trainees can prepare and exploit the potential learning benefits associated with school practice.

11.1 School Practice

School practice, also known as *teaching practicum* in some teacher educational contexts, is a formally organized and well-structured educational activity that teacher students are expected to successfully complete as part of their teacher professional training. It is purposely organized to provide teacher trainees with a basis and platform to practice teaching in a real context, namely the classroom. In so doing, trainees are expected to develop into effective practicing teachers, demonstrate the attained pedagogical and didactical competences, and to display

F. Kayima, *Preparing for Chemistry Teaching*, Springer Texts in Education,
https://doi.org/10.1007/978-3-031-97122-8_11

their progress in developing as professional teachers. It is through school practice (practicum) that trainees get a chance to put their personal theories about teaching to test (Kayima, 2022), a process which triggers and results in transitioned thinking about the nature and practice of teaching.

The way school practice or teaching practicum is organized and structured can be different from one context to another, but the overall aims and expectations are common across different educational contexts. In this chapter, the focus is on learning in and from practice. I consider and discuss the potential contributions of school practice towards a trainee's professional development into a competent teacher practitioner, the nature of teacher knowledge and skills attained through practice, what teaching perspectives the trainee comes with into school practice and what teaching perspectives, knowledge and skills a trainee leaves with at the end of school practice. The goal is to provide trainees and teacher educators with a background regarding the central learning elements of this teacher training component and how they can prepare and be ready to exploit the potential learning benefits associated with school practice.

11.2 The Teacher Trainee's Knowledge Base—Before School Practice

As earlier mentioned, school practice or teaching practicum is organized such that trainees get the opportunity to practice teaching in real contexts. It presupposes that the candidate has acquired sufficient professional training that includes taking specific courses in the relevant teaching subjects, as well as didactics and pedagogical courses. In other words, the candidate must possess satisfactory knowledge of the subject matter (content knowledge), knowledge about teaching generally (pedagogical content knowledge), as well as knowledge about the teaching and learning that relates to the subject to be taught (subject specific pedagogical content knowledge). For simplicity, let us use the term "professional training" to coin all these three teacher knowledge elements, since professional training accounts for training in the subject matter (chemistry in our case) and training in teaching pedagogies and subject didactics (e.g., theories and perspectives about teaching and learning).

Moreover, in addition to obtaining professional training as described above, the teacher trainee carries with him/her an educational history—a set of experiences about teaching and learning accumulated over time as a student throughout the so many years of schooling. Among these experiences, the trainee has time and again observed, experienced, and has been taught by several different teachers using different teaching methods and approaches, stretching from very authoritative—monologic teacher controlled, to more open- dialogic student centered instructionally organized settings. Over the years, these educational histories and experiences about teaching shape the trainee's thinking about teaching and thereby contribute to a trainee's strong *belief system* (Bramald et al., 1995),—a set of principles, assumptions, values, attitudes, perceptions and convictions about teaching, learning, students, and the educational system, or what Kelchtermans (1993, 2009);

Vanassche and Kelchtermans (2014) among others, referred to as the *teacher's personal interpretive framework*—a set of "cognitions and beliefs that operates as a lens through which teacher educators perceive their job situations, give meaning to, and act in them." (Vanassche & Kelchtermans, 2014, p. 118).

Note that a teacher's or trainee's belief system and hence his or her personal interpretive framework is not only shaped by his/her educational histories and experiences but rather a combination of several encountered experiences, including learning experiences from professional training. Going into school practice therefore, the trainee takes with him/her a teaching knowledge and skills' base comprising (i) his/her belief system as described above and (ii) his/her professional training—knowledge of the subject matter, knowledge about teaching and learning, along with other learned and attained skills and competencies relevant to teaching contexts including knowledge about teaching with technology (Fig. 11.1). Moreover, as you read the above and reflect on your own experiences, you will realize that you also hold ideas about teaching and learning. We all do. These are continuously being shaped, reshaped, reinforced or strengthened over time whenever we come face to face with the reality of the teaching and learning complexity.

As you set out to start your school practice, your educational institution has marked some expectations for you. For example, the expectation that with your subject, pedagogical and didactics knowledge competencies, you will be able to

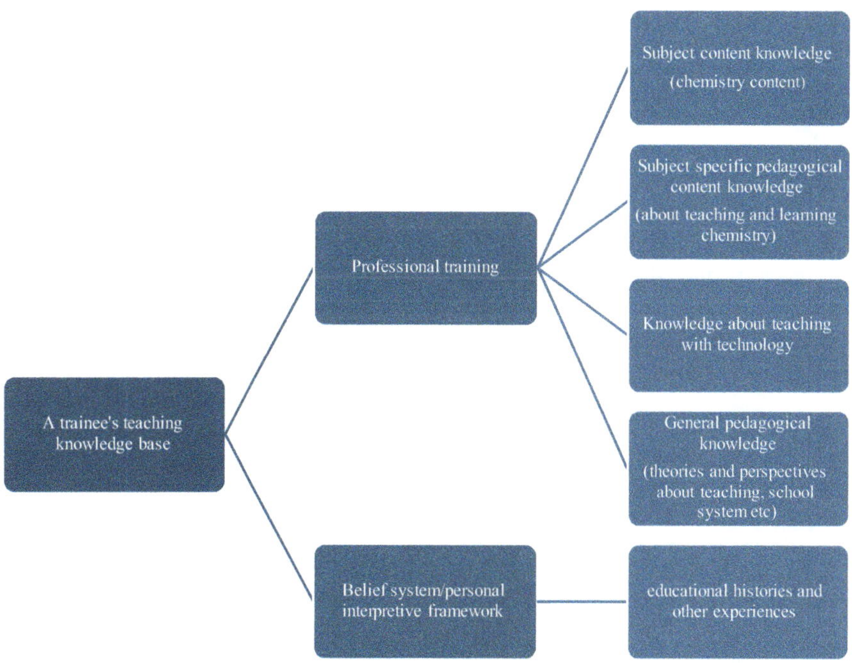

Fig. 11.1 Elements of a trainee's knowledge base

develop a teaching plan, implement your teaching plan using varied and student-centered instructional strategies, invoke a dialogic classroom discourse, display creativity, innovativeness and the ability to integrate and use relevant technologies in your teaching as well as manage and organize a classroom environment where most (if not all) students' needs are considered. Moreover, you are expected to master your teaching subject well and can break down concepts and present them to your students in a form that is digestible by the students. It is also expected that you will be able to plan and implement assessment strategies aimed at assessment for learning (formative assessment) and assessment of learning (summative assessment). As shown in Fig. 11.2, trainees are often expected to account for at least three questions when planning for their teaching, namely (i) what is it that you want to teach your students? (ii) how do you intend to teach that you wish, desire or have decided to teach? And (iii) how will the learning that has taken place be measured or how do you plan to establish the attainment of the intended learning?

There is no doubt about the fact that your teaching will be substantially influenced by your inherent belief system or your interpretive framework, and that you are poised to draw on those educational histories and experiences that you cherished or admired as good teaching experiences and identified yourself with, to plan and thereby emulate those practices. This might serve as an entry point for especially those who are very anxious and insecure, given that it's your first time being in school practice. Fortunately, the goal of this chapter is to enlighten you about the kind of knowledge and skills you have accumulated and the potential competences you exhibit, with which you are to confront your teaching practice with full confidence. The knowledge base that you have as summarized in Fig. 11.1 is to be harnessed—that is, collected, aligned and controlled for effective use. You will implicitly be working towards reconciling your belief system (– ideas and thinking about teaching) with perspectives, knowledge and skills from your professional training. This process which comes naturally is propelled by your willingness to make yourself aware of your inherently held ideas about teaching and assessing these against what you have learnt from teacher education training.

Understanding the existence of the two constituent elements of your teaching knowledge base will make it easier for you when you start planning for your teaching and when in actual teaching situations. You should not at any one moment doubt your own ideas about how teaching should look like or your thoughts about how to deal with different teaching situations. Having doubts about one's potential and thinking is not the same as putting your ideas and thinking to test, to find out whether they effectively serve the purpose or not. The goal should not be about falsifying your ideas but to align your thinking with what you have learnt in teacher training, and then you will naturally discard those ideas and adopt new ways of dealing with teaching situations as you continue to experience the realities of actual teaching situations. It is thus important to be aware and to understand that your educational histories and experiences are an important part of the interpretive framework that informs your teaching decisions (actions, behavior and other classroom performances). Having this awareness is key to opening-up for learning in and with practice. Very soon and as you continue with your school practice, you

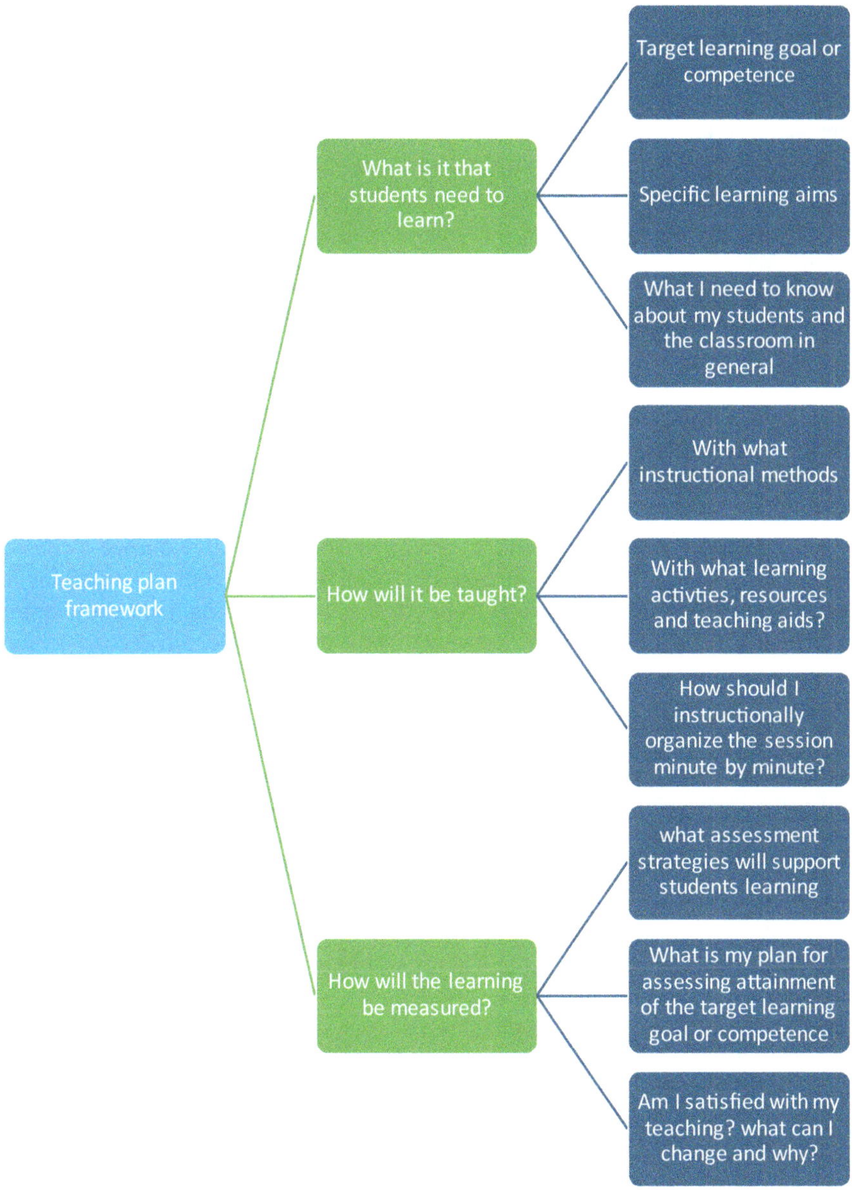

Fig. 11.2 A teaching plan framework comprising elements to be considered when you plan a lesson

will start to experience and visualize a relationship between your inherent ideas about teaching, your real-time experiences about teaching and the knowledge and ideas from professional training that have been operationalized or used in your teaching. There will be a form of exchange of information between your belief system, what you experience in real-time teaching situations, and how the two relate with perspectives, knowledge and skills from your teacher training. In other words, you are continuously drawing on both your belief system and your knowledge from teacher training (professional training) to make in-class timely decisions. The ability to make in-class and timely decisions during real-time teaching situations is often enabled by one's ability to quickly perform a reflection-in-action as part of what is called reflective practicing (Kayima, 2022) which is elaborated in the next section.

11.3 Learning in and from Practice—During School Practice

About 26 years ago (as of May 2025), Ball and Cohen (1999) published a chapter about "developing practice, developing practitioners", where they describe important elements about learning in and from practice, accounts which are still relevant to date and going forward. It is one thing to have the knowledge and competences needed to be able to plan for teaching such as the knowledge base described in the previous section. However, the requirements in real teaching situations come with other kinds of knowledge, skills and competencies, a large part of which are skills and knowledge that are learned in and from practice. This latter knowledge and skills constitute what was described in Chapter 2 as a teacher's knowledge for teaching acquired with and in practice. Both novice and expert teachers are expected to learn how to learn in contexts of their respective workplaces. In addition to the knowledge base described above (e.g.; Fig. 11.1), a teacher is expected to have the ability to "think on your feet" (Barnett & Hodson, 2001), deal with real time teaching situations, including making real-time adjustments to your teaching, attending to students varied needs, dealing with undesired and desired students' behavior, responding to satisfactions and dissatisfactions, while ensuring that the attainment of the intended learning goals and targets is realized.

While at teacher training, teacher educators facilitating your professional training do not directly tackle or teach this kind of knowledge and skills. In as much as they would wish to do so, the complex and practical nature of this type of teacher knowledge and abilities make it difficult to characterize and directly address during professional training. What teacher educators often do is talk about the need for this kind of knowledge and abilities, citing some few contextualized case situations and examples, and then talk about the challenges that await you in real teaching contexts as a way of preparing you to be ready to learn in and from practice. Learning in and from practice implies that you are prepared and have the capability to learn about students, anticipate different ways students can think, behave and react in different situations, model different ways to deal with different real-time teaching situations, learn to control yourself emotionally under stressful situations, the ability to reflect both in and on action (e.g.; Schön, 1983, 1987), take a step back, analyze the situation and devise a timely but appropriate course of action, manage a timely and real-time self-evaluation, pose questions to yourself about what is working and not, and find answers to these questions as you adapt to the changing classroom context. As Ball and Cohen (1999) expressed about 26 years ago, teaching will always require improvisation, conjecturing, experimenting and assessing, and school practice gives you the best opportunity to develop into a professional practitioner. It offers you the chance to experience what it means to adapt and develop practice.

11.4 Meeting with Your Mentor—The Teacher Educator or the Expert Teacher

During my over 15 years as a science teacher educator, I have participated in several school practice visits, to meet and observe my students (teacher trainees) in real teaching contexts. School practice visits by teacher educators mainly serve two things, first, to observe the student in a real teaching situation and provide an assessment regarding to his or her performance in the observed teaching capacity. Second, to provide productive feedback to the student in question, the kind of feedback needed to support the student's development in the journey of becoming a professional practitioner. In many cases, this kind of feedback concerns teaching actions or performances that the student has carried out well (very well or excellently) and those observed actions, performances or activities where improvement is needed. Educators are expected to provide productive and concrete feedback with supporting examples wherever possible, to guide the student into what needs to be done to improve, or to guide them in those areas where more attention is needed. Moreover, such feedback that entails giving commentaries on the observed specific performances as well as suggestions to improve, is more productive if the practicing student is provided with explanations or a justification for why a certain teaching action or performance should be carried out in a manner different from how the practicing student has carried it out, etc. This explanation or justification provides a basis for reflection, where the practicing student takes a step back to reflect about his/her own thinking or implemented idea and analyses his/her own thinking/implemented idea with respect to what is being suggested by the educator. Such a reflection is poised to trigger an alignment process involving the practicing student examining his/her own interpretive framework (or the belief system) and perspectives from educational training which are being reinforced by the visiting teacher educator.

Moreover, the meeting between the practicing student (teacher trainee) and his/her teacher educator from the training institution, if well used, substantially enhances the trainee's development of the kind of teacher knowledge that is obtained in and from practice as described in the section above and in Chapter 2. The educator's aim is to trigger a reflective process through providing productive and concrete feedback, as well as moderating a discussion about the observed actions or performances vis-à-vis the planned actions and performances (Kayima, 2022). As mentioned in the earlier section, real-time teaching involves improvisation, making timely decisions in response to real-time changes in the teacher-student discourse. Whereas it is possible for the teacher trainee to reflect on his/her practice at the end of the lesson, capturing moments of reflection-in-action during real-time teaching can be very difficult for the teacher trainee given that he/she is less experienced. Actions and decisions are rapidly taken during real-time teaching as the trainee works towards the attainment of the set learning targets. Having the educator observe these real-time incidents provides the trainee with the needed feedback concerning how she/he reacted in real-time situations, what courses of action were or have been taken, and how best similar incidents would be dealt with in future or in the next teaching.

That said, during my many school practice visits over the years, I have experienced a few situations where the lessons I have had to observe were fixed, superficially organized, and superficially carried out. This practice is unproductive and does not serve the important aim of school practice. These few fixed lessons that I observed were not naturally aligned with the classroom timetable or the current teaching plan within that teaching subject but had been only arranged to serve the purpose of my visit. In the worst-case scenario, even the classroom students themselves in some of these lessons appeared confused, unprepared for the lesson, some even without relevant or proper books or materials for the subject in question. This is obviously a bad learning experience for student teachers. Organizing and arranging school practice visits superficially by fixing lessons, lobbying for students to attend the classroom session for the sake of simply being observed by the visiting educator, does not serve the learning that I have labored to describe in earlier sections. Perhaps the trainee in these cases is more interested in getting his/her school practice training component approved (passed or graded with a pass) than having interest in the actual learning gained from the school practice experience. The expectation therefore is that you, the teacher trainee, will strive to engage those responsible for scheduling the teaching in the host school, to allow for a proper integration of your teaching practice episodes into the normal teaching routine.

11.5 Develop Assertions About Teaching and Transform Your Thinking

To conclude, it should be clear that expectations from the school practice training component are much more than simply requiring trainees to develop and implement a teaching session that considers several elements emphasized in teacher education, such as students' engagement and activation, instructional methods, resources and other aids, etc. In addition to this expectation, it is without doubt that school practice is also an arena for testing ideas and inherent thinking about teaching. Strong claims and arguments have been made in the past, in several empirical educational study reports about how resistant teacher trainees are when it comes to changing their held views (belief system) about teaching. A trainee's belief system has been thus positioned in the past as one of the impeding factors to a trainee's professional development (see Bramald et al., 1995 for these sources). To think that you will resist change in this fast moving and technologically driven society sounds like deserting your reasonability to contribute to society by embracing the current threshold demands for one to be able to contribute. Belief systems or the trainee's personal interpretive framework, his or her educational histories and experiences, and his/her professional training, are all important and they all play an important role in making and shaping a trainee into a professional practitioner. Learning today demands that we know and understand what we know, how we think, and whenever possible, how we have come to think the way we think and operate. Once we are able to define ourselves, we open up for seeking to know that which we don't know or understand, we embrace new challenges by posing questions to both ourselves and to others who think or act differently, about actions, performances, knowledge, technologies, and skills that we have not mastered, we pose and seek answers to questions about the potential benefits that come with embracing a different thinking or perspective, new ideas, actions and performances. We can no longer subscribe to or take pride in the idea of being resistant to changes since this is a luxury we cannot afford now in this century with all its complex challenges and problems.

Consequently, when we go into school practice as teacher trainees, we take with us a knowledge base, a set comprising knowledge from professional training and our belief system (see Fig. 11.3). This is our toolbox as teacher trainees and along with a list of expectations from our respective training institutions, we start our school practice journey. We draw on this pool of knowledge, namely our knowledge base, to develop, plan and implement classroom teaching sessions. The goal is not to stick to the practice as we know it, but to carry out the practice as we know it, reflect on the carried-out practice, learn from our failures and successes, from the things that have worked and from those that have not. Each time we remain open to learning from every interaction that we participate in, we pave way for re-analyzing our belief system, adjust where necessary, and thereby transform our thinking. Through testing our ideas and thinking in real teaching contexts, we refine our initial assertations and transform our personal theories about how teaching should be instructionally organized, about how content should be sequenced, about how to react in complex situations, and so on. Moreover, we get to test and assess the applicability of didactic frameworks and other learning perspectives, embracing them and using them as conceptual lenses that guide our classroom performances and courses of action. This is what school practice is about. It's where you get to understand and learn about the dynamics and complexities consisted in ***art of teaching*** while you are at the very center of it. After school practice, your knowledge base for teaching is refined, your thinking has changed in one way or another, and your perspective about the complexity of teaching has widened and thus you are more relieved and secure than you were and felt before.

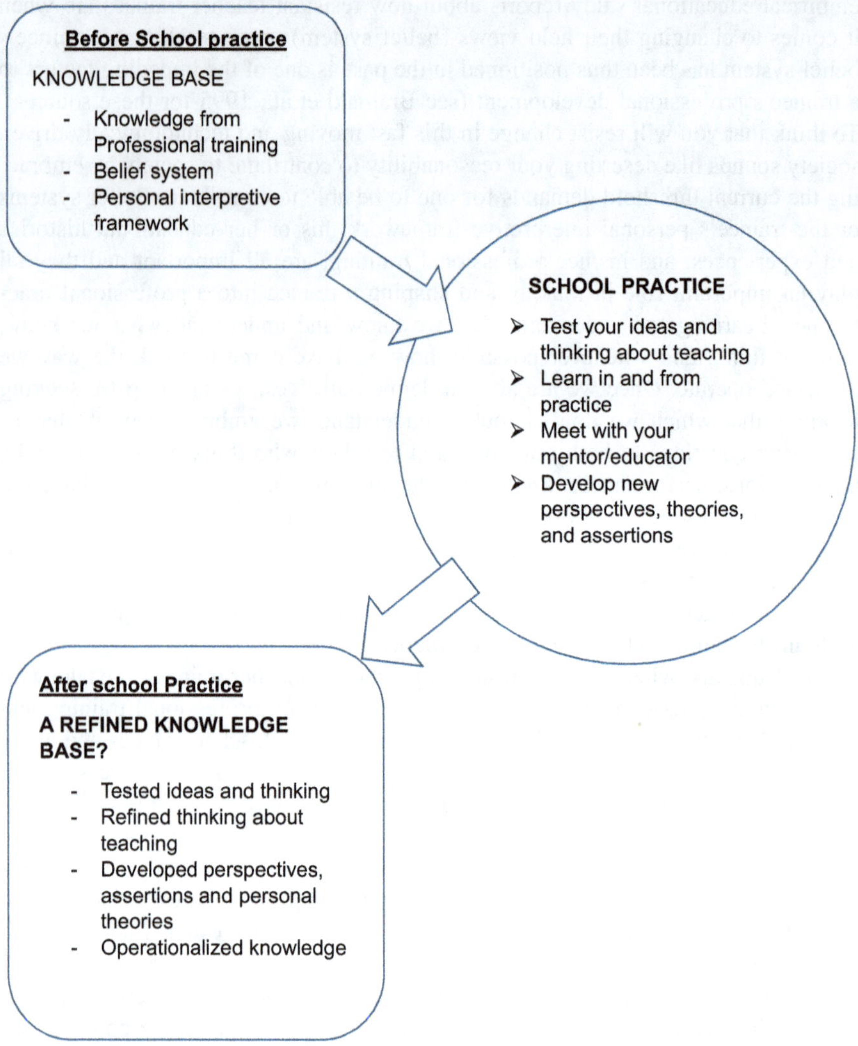

Fig. 11.3 Towards a refined knowledge base for teaching

References

Ball, D. L., & Cohen, D. K. (1999). Developing practice, developing practitioners: Toward a practice-based theory of professional education. In G. Sykes & L. Darling-Hammond (Eds.), *Teaching as the learning profession: Handbook of policy and practice* (Vol. 1, pp. 3–32). Jossey Bass.

Barnett, J., & Hodson, D. (2001). Pedagogical context knowledge: Toward a fuller understanding of what good science teachers know. *Science Education, 85*(4), 426–453. https://doi.org/10.1002/sce.1017

Bramald, R., Hardman, F., & Leat, D. (1995). Initial teacher trainees and their views of teaching and learning. *Teaching and Teacher Education, 11*(1), 23-31. https://doi.org/10.1016/0742-051X(94)E0009-T

Kayima, F. (2022). The role of reflective practice in mediating development of pre-service science teachers' professional and classroom knowledge. *Interdisciplinary Journal of Environmental and Science Education, 18*(1), e2262. https://doi.org/10.21601/ijese/11364

Kelchtermans, G. (1993). Getting the story, understanding the lives: From career stories to teachers' professional development. *Teaching and Teacher Education, 9*(5), 443-456. https://doi.org/10.1016/0742-051X(93)90029-G

Kelchtermans, G. (2009). Who I am in how I teach is the message: Self-understanding, vulnerability and reflection. *Teachers and Teaching, 15*(2), 257–272. https://doi.org/10.1080/13540600902875332

Schön, D. A. (1983). *The reflective practioner*. Basic Books.

Schön, D. A. (1987). *Educating the reflective practitioner*. Jossey-Bass.

Vanassche, E., & Kelchtermans, G. (2014). Teacher educators' professionalism in practice: Positioning theory and personal interpretative framework. *Teaching and Teacher Education, 44*, 117–127. https://doi.org/10.1016/j.tate.2014.08.006

www.ingramcontent.com/pod-product-compliance
Lightning Source LLC
Chambersburg PA
CBHW050233180825
31264CB00003B/20
9783031971211